ADVANCED SOLID MECHANICS:
SIMPLIFIED THEORY

ADVANCED SOLID MECHANICS: *SIMPLIFIED THEORY*

Farzad Hejazi and Tan Kar Chun

CRC Press
Taylor & Francis Group
Boca Raton London New York

CRC Press is an imprint of the
Taylor & Francis Group, an **informa** business

First edition published 2021
by CRC Press
6000 Broken Sound Parkway NW, Suite 300, Boca Raton, FL 33487-2742

and by CRC Press
2 Park Square, Milton Park, Abingdon, Oxon, OX14 4RN

Library of Congress Cataloging-in-Publication Data

Names: Hejazi, Farzad, author. | Tan, Kar Chun, author.
Title: Advanced solid mechanics : simplified theory / Farzad Hejazi and Tan Kar Chun.
Description: First edition. | Boca Raton : CRC Press, 2021. | Includes bibliographical
references and index.
Identifiers: LCCN 2020049289 (print) | LCCN 2020049290 (ebook) | ISBN
9780367705398 (hardback) | ISBN 9780367705381 (paperback) | ISBN 9781003146827
(ebook)
Subjects: LCSH: Mechanics, Analytic. | Solids--Mathematical models.
Classification: LCC QA807 .H425 2021 (print) | LCC QA807 (ebook) | DDC 531/.2--dc23
LC record available at https://lccn.loc.gov/2020049289
LC ebook record available at https://lccn.loc.gov/2020049290

ISBN: 978-0-367-70539-8 (hbk)
ISBN: 978-0-367-70538-1 (pbk)
ISBN: 978-1-003-14682-7 (ebk)

Typeset in Times
by MPS Limited, Dehradun

Contents

List of Figures

List of Tables

Authors

Farzad Hejazi is an Associate Professor and Research Coordinator at the Department of Civil Engineering, Faculty of Engineering, University Putra Malaysia (UPM), where he has also been an innovation champion since 2013 and a member of the Housing Research Center's management committee. He received his PhD in Structural Engineering from the University Putra Malaysia in 2011 and worked as a postdoctoral fellow until 2012. In addition, he teaches PhD students in structural engineering fields such as the finite element method, structural dynamics, advanced solid mechanics, advanced structural analysis and earthquake resistance structure.

Dr Hejazi manages and supervises a research team, consisting of 20 PhD students and 8 master's students, which is involved in various high impact research and industry projects funded by the Ministry of Higher Education Malaysia, Ministry of Science, Technology and Innovation, PlaTCOM Venture Malaysian Government Agency, University Putra Malaysia and industrial companies and has led to 15 patents being filed in the United States, Japan, Germany, Canada, New Zealand and Malaysia. Four of his patents are related to vibration dissipation devices already licensed to industry for mass production and implemented in various construction projects, such as bridges and other structures.

Tan Kar Chun holds a first-class honours degree in Civil Engineering from the University Putra Malaysia (UPM), and is currently pursuing his PhD in Structural Engineering at UPM. He has been a consultant in the civil and structural engineering industry for the past five years, and during that time he has been involved in multiple projects at various scales: from a 1,400-acre township development to a 50-storey high-rise building.

1 Introduction

1.1 MATTER

Any substance that has mass and volume, i.e. it occupies spaces, is matter. Matter is made up of discrete particles and can be categorised into three common states: solid, liquid and gas, based on the characteristics of its particles.

Solid is a state of matter with highest rigidity. A strong force of attraction between solid particles and the low molecular energy that restrains the particles' movement endow solids the ability to hold on to their own shape. Inversely, a gas particle possesses high energy, while the attraction force between the particles is too weak to hold them in place.

Regardless of the strength of the force of attraction, particles can never fit perfectly with one another. Gaps and discontinuities are present between particles, and, as a result, the properties of a substance vary at different locations on it at the microscopic level due to different arrangements of particles, as shown in Fig. 1.1.

For engineering purposes, however, the microscopic discontinuities and resultant variations are neglected by considering the substance as a continuum. A continuum is an idealised matter the particles of which are continuously distributed and fill the entire space it occupies. Thus, in such an idealised matter, discontinuities are completely removed. This matter can be divided into infinitesimal elements, and all of them can

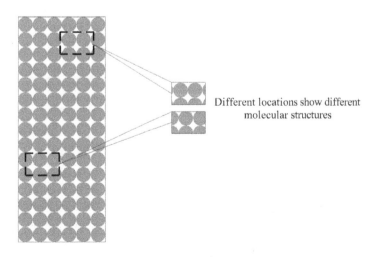

Different locations show different molecular structures

FIGURE 1.1 Actual solid at the microscopic level.

exhibit the same properties. Under this condition, the properties of the matter are expressed as a continuous function of space and time. This is demonstrated in Fig. 1.2.

A matter can be categorised into fluid or solid. A fluid is a substance that will deform continually when shear stress is applied on it. In other words, a fluid does not have shear resistance. By this definition, liquid and gas are both categorised as fluids. By contrast, solids can resist shear with their attraction force and hence do not deform continually when subjected to shear stress.

The study of a solid's behaviour, e.g. deformation and stress distribution in both linear and non-linear manner under external forces, is known as solid mechanics, and such a study for fluids is known as fluid mechanics. Both studies are grouped as continuum mechanics, since the subject of study, i.e. the substance, is always considered a continuum.

1.2   LOCATION AS A FACTOR AFFECTING MATERIAL PROPERTIES

By assuming a material as continuum, the variation in properties at different locations on the material is solely dependent on its uniformity. The uniformity of a material is defined by the consistency in term of its composition at any point.

1.2.1   HETEROGENEOUS MATERIAL

A heterogeneous material shows a distinctive composition at the microscopic level, at any location on it. Since all constituent materials are not evenly distributed over the material, its properties are said to be dependent on the location on the material.

An example for heterogeneous material is reinforced concrete, which is a composite consisting of two main construction materials: concrete and steel.

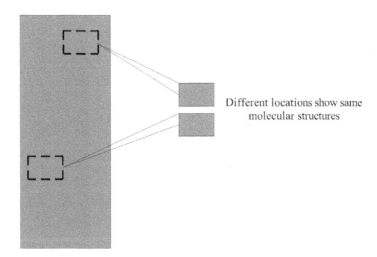

Different locations show same molecular structures

FIGURE 1.2   Continuum at the microscopic level.

The dominant constituent material, concrete, shows good compression but poor tension. On the other hand, steel shows good tension, but its strength declines tremendously after being subjected to very high temperature and corrosion.

The combination of concrete and steel, i.e. reinforced concrete, is an economical solution to improving the structural member's resistance to compression, tension, bending and shear.

For example, when load is applied on the top of a beam, a sagging moment is induced. The top of the beam is subjected to compression, while the bottom is subjected to tension. The primary reinforcing steel bars are placed at the bottom of the beam to help resist the tension. Concrete is casted around the steel bars to hold them in place and provide protection against high temperature and corrosion.

Consider that the tension is applied at two different locations on a reinforced concrete, as shown in Fig. 1.3 below. Due to the difference in composition at different locations, the material's behaviour varies at the point that consists of concrete only; the deformation there is significantly higher than that at the point that consists of both concrete and steel.

1.2.2 HOMOGENEOUS MATERIAL

An ideal homogeneous material shows a uniform composition at the microscopic level, at any location on it. Since all constituent materials are well-distributed over the material, its properties are said to be independent of the location on the material.

An example of a homogeneous material is stainless steel. Its constituent materials are iron ore, chromium, silicon, nickel, carbon, manganese and nitrogen. The first step in manufacturing stainless steel is heating the constituent materials, melting them and letting them mix. The process is important to ensure the final product is homogeneous. Fig. 1.4 shows the steel manufacturing process as described above.

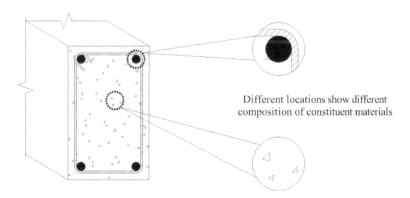

Different locations show different composition of constituent materials

FIGURE 1.3   Reinforced concrete as a heterogeneous material.

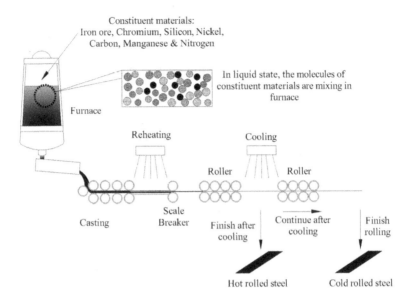

FIGURE 1.4   Hot-rolled and cold-rolled steel manufacturing process.

1.3   ORIENTATION AS A FACTOR AFFECTING MATERIAL PROPERTIES

A material's behaviour also depends on its orientation with respect to the applied force. Ideally, a material is assumed isotropic to simplify engineering analysis and design. However, this is nearly impossible in the real world, as a perfect homogeneous material is impossible to be created.

Due to the imperfect uniformity in term of composition, the material will react differently to the same magnitude of force that acts along different directions.

1.3.1   ANISOTROPIC MATERIAL

An anisotropic material shows six different mechanical properties when force is acting in six different directions along three mutually orthogonal axes, i.e. principal axes. In other words, its properties are dependent on the orientation of the material.

A composite is usually anisotropic. It is created by combining two or more constituent materials with different properties. The final material is created by arranging these constituent materials in either a specific or a vnon-specific order without breaking the arrangement of their particles. This makes the final material anisotropic, because the properties of each constituent material are only present throughout the space that such a constituent material occupies.

An example of a composite is fibre-reinforced concrete. In a fibre-reinforced concrete block, concrete provides principal resistance to compression, while fibre, e.g. synthetic fibre, provides principal resistance to tension. Fibre is an anisotropic material. It shows highest resistance to tension only when the force is acting along its longitudinal axis. Therefore, the orientation and arrangement of fibre directly affect the tensile strength of the concrete.

In a fibre-reinforced concrete block, the arrangement and orientation of constituent materials, aggregates, sand, cement (components of concrete) and fibres are always arbitrary in every direction. This causes the tensile strength of fibre-reinforced concrete to vary in different directions, as shown in Fig. 1.5.

1.3.2 ORTHOTROPIC MATERIAL

An orthotropic material shows three mechanical properties along three principal axes. Therefore, the properties are dependent on the orientation of the material.

An example for orthotropic material is timber. Timber originates from trees. The most unique characteristic of a tree is the presence of annual rings that surround its pith. For every year the tree lives, a new ring will be developed as the outermost layer of its trunk. This layer of annual rings is known as grain in timber, and it is the reason that makes timber behave as an orthotropic material.

The compressive and tensile strength of the timber varies along each principal axes of timber, i.e. radial, tangential and longitudinal are different, due to the arrangement and pattern of its grain pattern. As shown in Fig. 1.6, forces acting along both 1-1 (radial direction) and 2-2 (tangential direction) axes are perpendicular to the grain, while the force acting along the 3-3 (longitudinal direction) axis is parallel to the grain.

When the force is perpendicular to the grain, discontinuities between the resisting members, i.e grains, present along the direction of the force. In this case, the grains cannot deploy their resistance to the force because the adhesion between the

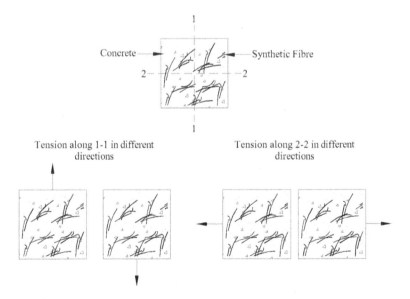

FIGURE 1.5 Fibre-reinforced concrete as an anisotropic material.

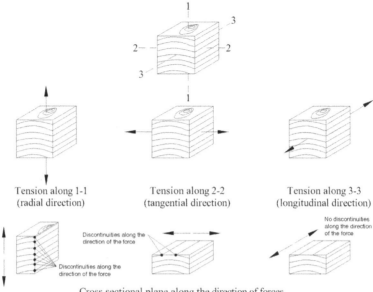

FIGURE 1.6   Timber as an orthotropic material.

grains is very weak and grains' joint failure will occur first. As a result, grain-splitting occurs at discontinuities and the block is fully destroyed.

When the force is parallel to the grain, no discontinuity presents along the direction of the force. The grains can deploy their full resistance to the force. Hence, the tensile and compressive strength of timber is very high compared to the case where the force is perpendicular to the grains.

1.3.3   ISOTROPIC MATERIAL

An isotropic material shows the same mechanical properties regardless of the direction of the force. Therefore, the properties are independent of the orientation of the material.

A material can achieve isotropy only if all its constituent materials are well-mixed. In such a mixture, the particles of all constituent materials are mixed together and the distinctive properties from each constituent material average each other. Therefore, an isotropic material must be homogeneous.

An example for isotropic material is stainless steel. By going through the manufacturing process as shown in Fig. 1.4, the properties of all constituent materials are averaging each other. The material will display the same behaviour regardless of the direction and location of the applied forces.

2 Stress

2.1 FORCE AND STRESS

Force can be described as two distinctive actions: push and pull. A body reacts to exerted force by changing its velocity based on the force's magnitude and direction. According to Newton's third law of motion, for every unit of force exerted by one body on another, an equal magnitude of force will be exerted back to it in the opposite direction. Take a restrained girder, as shown in Fig. 2.1, as an example, where supports *1* and *2* are, in fact, two bodies in contact with the girder.

When force is exerted on the girder, it is transmitted throughout the body. When the force is transferred to the point of contact with any of the restrain, Newton's third law of motion will come into play and a reaction force will be exerted by the support to the girder.

Force is transmitted throughout the body via particles. At the microscopic level, particles will change their velocity from zero (at rest) to a certain value. The moving particles will fill the void between them and because of the attraction force, the nearby particles are pushed or pulled as well. In short, particles will experience an internally developed force in every direction.

By cutting the solid and inspecting the sectional plane, one can find such internal forces acting on that plane. The concept of stress, which is the average of the resultant internal forces distributed over that sectional plane, is introduced.

In engineering, two main types of force are concerned: normal force and shear force. Therefore, normal stress and shear stress are two fundamental types of stress discussed in solid mechanics. Normal stress is developed by a normal force acting perpendicularly to a plane (Fig. 2.2) while shear stress is developed by a shear force acting parallelly to a plane (Fig. 2.3).

$$Normal\ stress,\ \sigma = \frac{F}{A} \tag{2.1}$$

where
  F is the normal force;
  A is the area of sectional plane.

where
  V is the shear force;
  A_v is the area of the shear plane.

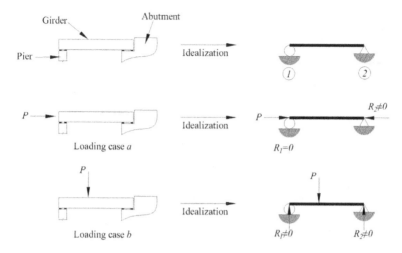

FIGURE 2.1   Girder subjected to point loads.

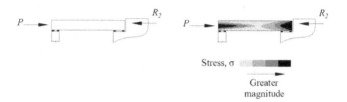

FIGURE 2.2   Girder subjected to normal force.

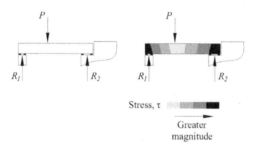

FIGURE 2.3   Girder subjected to shear force.

2.2   COMPONENTS OF STRESS

Solid mechanics studies the stresses and strains at any point on a body, usually illustrated using an infinitesimal cube enveloping the point. Since solids are three dimensional, we can define three mutually orthogonal axes by setting that point as centre. The stress and strain at this point can be categorised based on their direction, with each of them acting along a certain axis. Without other axes to be compared,

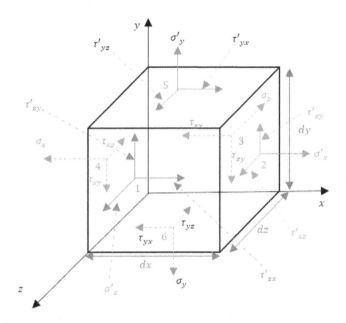

FIGURE 2.4   Stress components at a point.

the defined axes can be assumed not to have any inclination. This will be our reference axes, or global axes, as indicated in Fig. 2.4.

Suppose a force is exerted along the x-axis. The developed stress acting along the x-axis is distributed over the plane yz, which is the plane normal to the x-axis. The resultant stress is normal stress along the x-axis, σ_x. Meanwhile, the stress acting along the y-axis is distributed over the face of the plane yz, which often results in a change in the plane's shape. The resulting stress is shear stress distributed along the y-axis due to the force acting parallelly on the normal plane of the x-axis, τ_{xy}. Similarly, τ_{xz} is known as the shear stress distributed along the z-axis due to the force acting parallelly on the normal plane of the x-axis. The notation for strain components can be interpreted in the same way.

Table 2.1 shows the generalised stress components on each face in all directions. All 18 unknown stress components need to be solved as all of them are independent of each other.

Shear stress is coupled. Then, by taking the resultant of normal stresses along each axis, the stress at a point is said to be defined completely by nine independent components (three normal and six shear components). The components of stress are as follows:

$$\sigma = \begin{bmatrix} \sigma_X & \tau_{xy} & \tau_{xz} \\ \tau_{yx} & \sigma_Y & \tau_{yz} \\ \tau_{zx} & \tau_{zy} & \sigma_Z \end{bmatrix} \tag{2.3}$$

where $\sigma_X = \sigma_x + \sigma'_x$, $\sigma_Y = \sigma_y + \sigma'_y$ and $\sigma_Z = \sigma_z + \sigma'_z$.

TABLE 2.1

Unknown stress components for generalised case

		Plane		
		xy-plane (1 & 3)	*yz*-plane (2 & 4)	*xz*-plane (5 & 6)
Direction	positive *x*	τ'_{zx}	σ'_x	τ'_{yx}
	negative *x*	τ_{zx}	σ_x	τ_{yx}
	positive *y*	τ'_{zy}	τ'_{xy}	σ'_y
	negative *y*	τ_{zy}	τ_{xy}	σ_y
	positive *z*	σ'_z	τ'_{xz}	τ'_{yz}
	negative *z*	σ_z	τ_{xz}	τ_{yz}

2.3   STRESS EQUILIBRIUM EQUATION

Stresses is a continuous function of the location in a body. Therefore, the stress at any point is interrelated with the stress at other points. When force is acting along the *x*-axis of a solid, normal stress σ_x is developed at the contacting surface.

The solid is a continuum and internal force is exerted by the contacting particles on the others. As a result, stress develops in the entire solid body rather than just the contacting point. Considering the rate of stress development per unit length along *x*-axis as $\frac{\partial \sigma_x}{\partial x}$, the increment in stress across the length dx is:

$$\Delta \sigma_x = \frac{\partial \sigma_x}{\partial x} dx$$

If the point on solid is at rest and in equilibrium, the stress developed along the *x*-axis due to the aforementioned external force and internal force will be balanced by a stress of the same magnitude but in a different direction. Therefore, σ'_x can be expressed in term of σ_x:

$$\sigma'_x = \sigma_x + \frac{\partial \sigma_x}{\partial x} dx \qquad (2.4)$$

Similarly, the following normal stress components can be defined based on Eq. (2.4):

$$\sigma'_y = \sigma_y + \frac{\partial \sigma_y}{\partial y} dy$$

$$\sigma'_z = \sigma_z + \frac{\partial \sigma_z}{\partial z} dz$$

Six shear stress components can be expressed in a similar fashion:

$$\tau'_{xy} = \tau_{xy} + \frac{\partial \tau_{xy}}{\partial x} dx \; \tau'_{yx} = \tau_{yx} + \frac{\partial \tau_{yx}}{\partial y} dy$$

$$\tau'_{xz} = \tau_{xz} + \frac{\partial \tau_{xz}}{\partial x} dx \; \tau'_{zx} = \tau_{zx} + \frac{\partial \tau_{zx}}{\partial z} dz$$

$$\tau'_{yz} = \tau_{yz} + \frac{\partial \tau_{yz}}{\partial y} dy \; \tau'_{zy} = \tau_{zy} + \frac{\partial \tau_{zy}}{\partial z} dz$$

A total of 18 stress components can be expressed in the above-derived forms. All these unknown stress components can now be determined by knowing only nine of them, as shown in Table 2.2.

Other than the stresses developed on six planes, body forces may exist too. Let f_x, f_y and f_z be the body forces (force per volume) acting along the x, y and z axes, as shown in Fig. 2.5.

Under static equilibrium, the summation of forces acting on an infinitesimal part of the solid, as shown in Fig. 2.5, along the x-axis is zero. The involved forces are shown in Fig. 2.6.

$$\overset{+}{\rightarrow} \sum F_x = 0$$

$$\left[\sigma_x + \frac{\partial \sigma_x}{\partial x} dx\right] dydz - \sigma_x dydz + \left[\tau_{yx} + \frac{\partial \tau_{yx}}{\partial y} dy\right] dxdz - \tau_{yx} dxdz + \left[\tau_{zx} + \frac{\partial \tau_{zx}}{\partial z} dz\right] dxdy$$
$$- \tau_{zx} dxdy + f_x \, dxdydz = 0$$

TABLE 2.2
Unknown stress components for solid in equilibrium

		Plane		
		xy-plane (1 & 3)	yz-plane (2 & 4)	xz-plane (5 & 6)
Direction	positive x	$\tau_{zx} + \frac{\partial \tau_{zx}}{\partial z} dz$	$\sigma_x + \frac{\partial \sigma_x}{\partial x} dx$	$\tau_{yx} + \frac{\partial \tau_{yx}}{\partial y} dy$
	negative x	τ_{zx}	σ_x	τ_{yx}
	positive y	$\tau_{zy} + \frac{\partial \tau_{zy}}{\partial z} dz$	$\tau_{xy} + \frac{\partial \tau_{xy}}{\partial x} dx$	$\sigma_y + \frac{\partial \sigma_y}{\partial y} dy$
	negative y	τ_{zy}	τ_{xy}	σ_y
	positive z	$\sigma_z + \frac{\partial \sigma_z}{\partial z} dz$	$\tau_{xz} + \frac{\partial \tau_{xz}}{\partial x} dx$	$\tau_{yz} + \frac{\partial \tau_{yz}}{\partial y} dyA; l$
	negative z	σ_z	τ_{xz}	τ_{yz}

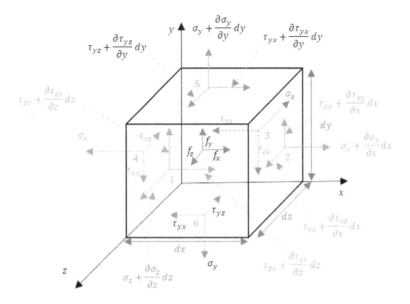

FIGURE 2.5 Components of stress in 3-D.

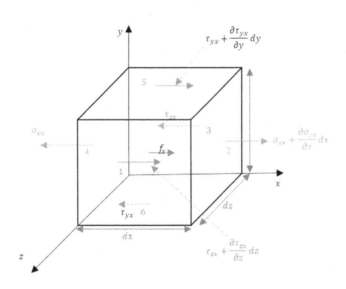

FIGURE 2.6 Stresses that act along x-axis.

$$\sigma_x dydz + \frac{\partial \sigma_x}{\partial x}dxdydz - \sigma_x dydz + \tau_{yx} dxdz + \frac{\partial \tau_{yx}}{\partial y}dxdydz - \tau_{yx} dxdz + \tau_{zx} dxdy$$

$$+ \frac{\partial \tau_{zx}}{\partial z}dxdydz - \tau_{zx} dxdy + f_x dxdydz = 0$$

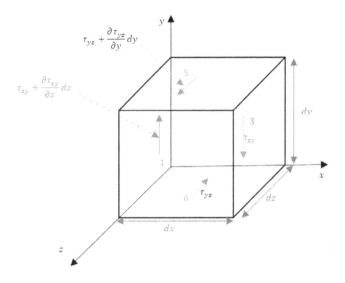

FIGURE 2.7 Stresses with the ability to cause rotation about x-axis.

$$\frac{\partial \sigma_x}{\partial x} + \frac{\partial \tau_{yx}}{\partial y} + \frac{\partial \tau_{zx}}{\partial z} + f_x = 0 \tag{2.5}$$

Similarly, equating the summation of forces along y and z axes to zero yields the following:

$$\frac{\partial \tau_{xy}}{\partial x} + \frac{\partial \sigma_y}{\partial y} + \frac{\partial \tau_{zy}}{\partial z} + f_y = 0 \tag{2.6}$$

$$\frac{\partial \tau_{xz}}{\partial x} + \frac{\partial \sigma_{yz}}{\partial y} + \frac{\partial \tau_z}{\partial z} + f_z = 0 \tag{2.7}$$

Eqs. (2.5), (2.6) and (2.7) are equilibrium equation for stresses considering body forces. Equating the summation of moment about the x-axis induced by forces (Fig. 2.5) to zero results in the following (Fig. 2.7):

$$\overset{\curvearrowright}{+}(\Sigma M_O)_x = 0$$

$$\left[\tau_{yz} + \overbrace{\frac{\partial \tau_{yz}}{\partial y}dy}^{\text{too small}}\right](dxdz)\left(\frac{dy}{2}\right) + \tau_{yz}(dxdz)\left(\frac{dy}{2}\right) - \left[\tau_{zy} + \underbrace{\frac{\partial \tau_{zy}}{\partial z}dz}_{\text{too small}}\right](dxdy)\left(\frac{dz}{2}\right)$$

$$- \tau_{zy}(dxdy)\left(\frac{dz}{2}\right) = 0$$

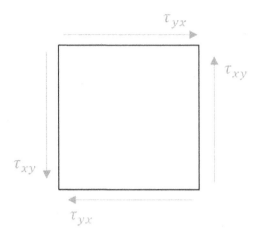

FIGURE 2.8 Complementary pairs of shear stresses.

Neglecting the 4th order products, because they are insignificant, gives the following:

$$\tau_{yz}\,dxdydz - \tau_{zy}\,dxdydz = 0$$

$$\tau_{yz} = \tau_{zy} \tag{2.8}$$

Similarly, the following relationships are produced by taking the summation of moment about the y and z axes as zero:

$$\tau_{xy} = \tau_{yx} \tag{2.9}$$

$$\tau_{zx} = \tau_{xz} \tag{2.10}$$

Eqs. (2.8), (2.9) and (2.10) show that for each pair of shear stress (say τ_{xy}) developed, a complementary shear stress (say τ_{yx}) of the same magnitude will also be developed on their adjacent face. These complementary shear stresses act in the direction that can stabilise the rotation that would be caused by the developed shear stress pair alone, as shown in Fig. 2.8.

Therefore, the stress components in Eq. (2.3) can be expressed as a symmetrical matrix:

$$\sigma = \begin{bmatrix} \sigma_X & \tau_{XY} & \tau_{XZ} \\ \tau_{XY} & \sigma_Y & \tau_{YZ} \\ \tau_{ZX} & \tau_{YZ} & \sigma_Z \end{bmatrix} \tag{2.11}$$

Now, the stress at a point is defined completely by six independent components (three normal and three shear) instead of nine independent components (three normal and six shear).

2.4   STRESS TRANSFORMATIONS

In real-life applications, force does not always act parallel to any of the global axes. To ease analysis, it can be expressed in terms of direction cosine after being divided into three components, along each of the global x, y and z directions.

After analysis, stresses and strains along global axes are determined. Using direction cosine, the stresses and strains can be changed into a resultant stress and strain that is inclined in all three axes. The plane that has this resultant stress as its normal is known as the oblique plane. The inclination of the oblique plane can be expressed in direction cosine as well.

Sometimes, rather than just referring to the global axes for every case, another set of mutually orthogonal axes can be defined by transforming the global axes with a particular inclination. This approach is usually used when there is no force exerted or developed along previously defined global axes. Such transformation requires least effort when the body is isotropic, where its mechanical properties are not dependent on its orientation, as shown in Fig. 2.9.

Consider a general 3-D vector, r , across three mutually orthogonal axes. Let α be the inclination angle between the vector and x-axis, β be the inclination angle between the vector and y-axis and γ be the inclination angle between the vector and z-axis, as shown in Fig. 2.10.

The trigonometric relationships between x, y, z and r are:

$$\cos \alpha = \frac{x}{r} = l$$
$$\cos \beta = \frac{y}{r} = m \qquad (2.12)$$
$$\cos \gamma = \frac{z}{r} = n$$

By applying the Pythagoras theorem, vector r can be expressed as:

$$r^2 = x^2 + y^2 + z^2$$

Divide all terms with r^2 to eliminate the term on the left-hand side and the following equation will be obtained:

$$\frac{r^2}{r^2} = \frac{x^2}{r^2} + \frac{y^2}{r^2} + \frac{z^2}{r^2}$$

Simplifying the equation above leads to

$$1 = \left(\frac{x}{r}\right)^2 + \left(\frac{y}{r}\right)^2 + \left(\frac{z}{r}\right)^2$$

Applying the relationship in Eq. (2.12) yields the following equation:

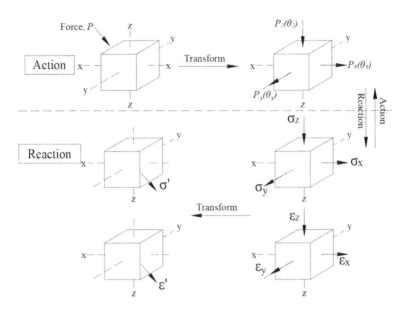

FIGURE 2.9   Transformation of stress and strain.

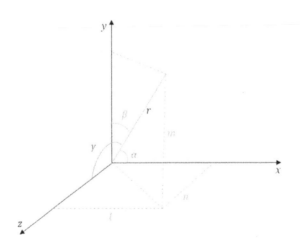

FIGURE 2.10   3-D vector across three mutually orthogonal axes.

$$1 = l^2 + m^2 + n^2 \tag{2.13}$$

Now, consider an infinitesimal tetrahedron ABCO, as shown in Fig. 2.11.
Let the faces of the tetrahedron be labelled as

$$ABC = A \ OAB = A_x \ OAC = A_y \ OBC = A_z$$

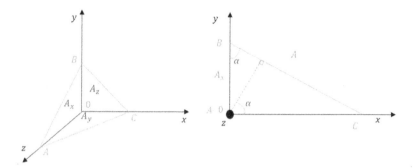

FIGURE 2.11 Tetrahedron in 3-D and xy-plane views.

Also, let α be the inclination angle between A and x-axis, β be the inclination angle between A and y-axis and γ be the inclination angle between A and z-axis. By cutting the tetrahedron at the xy-plane, angle OAC will be identified as α by applying the triangle exterior angle theorem. Therefore, the following can be obtained:

$$\cos \alpha = \frac{A_x}{A} = l \tag{2.14}$$

By expressing A_x in term of A gives

$$A_x = Al \tag{2.15}$$

Similarly, for both yz and xz planes, the trigonometric correlation can be expressed as

$$\cos \beta = \frac{A_y}{A} = m, \ A_y = Am \tag{2.16}$$

$$\cos \gamma = \frac{A_z}{A} = n, \ A_z = An \tag{2.17}$$

The forces on each plane are broken down into x, y and z components as shown in Fig. 2.12. Considering that the forces acting on each face of tetrahedron in x-direction are in equilibrium leads to

$$\sum F_x(+\rightarrow) = 0$$

$$0 = P_x A - \sigma_x A_x - \tau_{yx} A_y - \tau_{zx} A_z$$

Expressing A_x, A_y and A_z in terms of A yields the following equation:

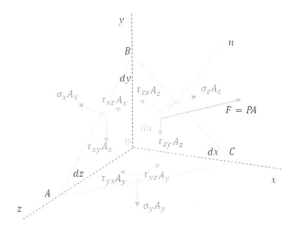

FIGURE 2.12 Breakdown of forces on the tetrahedron.

$$0 = P_x A - \sigma_x Al - \tau_{yx} Am - \tau_{zx} An$$

Rearranging the equation above produces the following expression:

$$P_x = \sigma_x l + \tau_{yx} m + \tau_{zx} n \qquad (2.18)$$

where P_x is the component of the surface force per unit area of face ABC, P, in x-direction.

Considering the forces acting on each face of tetrahedron in the y-direction to be in equilibrium, the following is obtained:

$$\sum F_y (+\uparrow) = 0$$

$$0 = P_y A - \tau_{xy} A_x - \sigma_y A_y - \tau_{zy} A_z$$

Expressing A_x, A_y and A_z in terms of A leads to

$$0 = P_y A - \tau_{xy} Al - \sigma_y Am - \tau_{zy} An.$$

Rearranging the equation above results in the following equation:

$$P_y = \tau_{xy} l + \sigma_y m + \tau_{zy} n \qquad (2.19)$$

where P_y is the component of the surface force per unit area of face ABC, P, in y-direction.

Similarly, the following equation is produced by considering the forces acting on each face of the tetrahedron in z-direction to be in equilibrium:

$$P_z = \tau_{xz}l + \tau_{yz}m + \sigma_z n \qquad (2.20)$$

where P_z is the component of the surface force per unit area of face ABC, P, in z-direction.

By arranging the relationships expressed in Eqs. (2.18), (2.19) and (2.20) in a matrix form, we get the following expression:

$$\begin{bmatrix} P_x \\ P_y \\ P_z \end{bmatrix} = \begin{bmatrix} \sigma_x & \tau_{yx} & \tau_{zx} \\ \tau_{xy} & \sigma_y & \tau_{zy} \\ \tau_{xz} & \tau_{yz} & \sigma_z \end{bmatrix} \begin{bmatrix} l \\ m \\ n \end{bmatrix}. \qquad (2.21)$$

The normal component of the resultant stress P, P_n (acting as normal to plane ABC), will be as follows:

$$P_n = P_x l + P_y m + P_z n. \qquad (2.22)$$

Substituting Eqs. (2.18), (2.19) and (2.20) into the equation above gives us the following expression:

$$P_n = (\sigma_x l + \tau_{yx}m + \tau_{zx}n)l + (\tau_{xy}l + \sigma_y m + \tau_{zy}n)m + (\tau_{xz}l + \tau_{yz}m + \sigma_z n)n.$$

Expanding the equation above leads to the following:

$$P_n = \sigma_x l^2 + \tau_{yx}lm + \tau_{zx}ln + \tau_{xy}lm + \sigma_y m^2 + \tau_{zy}mn + \tau_{xz}ln + \tau_{yz}mn + \sigma_z n^2.$$

Applying the relationships in Eqs. (2.8), (2.9) and (2.10) in the equation above gives us the following equation:

$$P_n = \sigma_x l^2 + \sigma_y m^2 + \sigma_z n^2 + 2\tau_{xy}lm + 2\tau_{yz}mn + 2\tau_{xz}ln.$$

Now, consider the tetrahedron rotated to the new coordinate system with three new mutually orthogonal axes namely x', y' and z'. The new axes are related to x, y and z through direction cosine, as shown in Fig. 2.13.

Table 2.3 shows the notation of direction cosines between new and old co-ordinate systems. For example, l_1 is the direction cosine between x and x'.

By referring to Eq. (2.13), the following relationships can be derived:

$$l_1^2 + m_1^2 + n_1^2 = 1,$$

$$l_2^2 + m_2^2 + n_2^2 = 1,$$

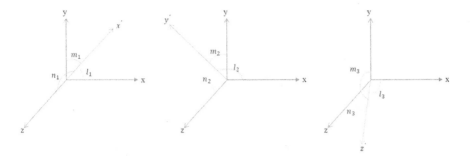

FIGURE 2.13 Position of new coordinate system relative to original coordinate system.

TABLE 2.3
Direction cosines between new and old coordinate systems

	x	y	z
x'	l_1	m_1	n_1
y'	l_2	m_2	n_2
z'	l_3	m_3	n_3

$$l_3^2 + m_3^2 + n_3^2 = 1.$$

The angle between two of the mutually orthogonal axes, ϕ , can be expressed in terms of their direction cosines. For example, the angle between x' and y' axes can be expressed as

$$\cos \phi = l_1 l_2 + m_1 m_2 + n_1 n_2.$$

Since both axes are mutually orthogonal, and $\cos \phi = 0$. Therefore, the equation above becomes

$$l_1 l_2 + m_1 m_2 + n_1 n_2 = 0.$$

The expression for other planes can be produced in a similar form as above:

$$l_2 l_3 + m_2 m_3 + n_2 n_3 = 0,$$

$$l_3 l_1 + m_3 m_1 + n_3 n_1 = 0.$$

Let the stresses be σ_{nx}, σ_{ny} and σ_{nz} along the x, y and z axes, and σ'_x, $\tau_{x'y'}$ and $\tau_{x'z'}$ along the x', y' and z' axes. The definitions are shown in Fig. 2.14.

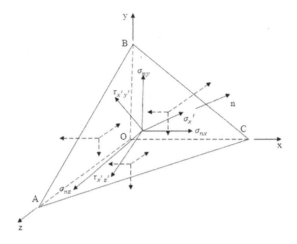

FIGURE 2.14 Stresses acting on plane ABC before and after rotation.

σ'_x can be expressed in terms of stresses (σ_{nx}, σ_{ny} and σ_{nz}) before rotation, in a form similar to Eq. (2.18) by applying direction cosines for x' with respect to x, y and z axes as per Table 2.3 (l_1, m_1 and n_1):

$$\sigma_{x'} = \sigma_{nx} l_1 + \sigma_{ny} m_1 + \sigma_{nz} n_1.$$

σ_{nx}, σ_{ny} and σ_{nz} are interchangeable with P_x, P_y and P_z because they, respectively, represent the same component of the surface force per unit area of face ABC, P, before rotation in x, y and z directions. Therefore, the equation above can be transformed to

$$\sigma_{x'} = P_x l_1 + P_y m_1 + P_z n_1.$$

Substituting Eqs. (2.18), (2.19) and (2.20) with relevant direction cosines into the equation above gives us the following:

$$\sigma_{x'} = (\sigma_x l_1 + \tau_{yx} m_1 + \tau_{zx} n_1) l_1 + (\tau_{xy} l_1 + \sigma_y m_1 + \tau_{zy} n_1) m_1 + (\tau_{xz} l_1 + \tau_{yz} m_1 + \sigma_z n_1) n_1.$$

Expanding the equation above leads to the following equation:

$$\sigma_{x'} = \sigma_x l_1^2 + \tau_{yx} l_1 m_1 + \tau_{zx} l_1 n_1 + \tau_{xy} l_1 m_1 + \sigma_y m_1^2 + \tau_{zy} m_1 n_1 + \tau_{xz} l_1 n_1 + \tau_{yz} m_1 n_1 + \sigma_z n_1^2.$$

Applying the relationships as per Eqs. (2.8), (2.9) and (2.10) in the equation above yields the following expression:

$$\sigma_{x'} = \sigma_x l_1^2 + \sigma_y m_1^2 + \sigma_z n_1^2 + 2\tau_{xy} l_1 m_1 + 2\tau_{yz} m_1 n_1 + 2\tau_{xz} l_1 n_1. \qquad (2.23)$$

The same principle for derivation can be applied for other components, say $\sigma_{y'}$:

$$\sigma_{y'} = \sigma_{nx} l_2 + \sigma_{ny} m_2 + \sigma_{nz} n_2,$$

$$\sigma_{y'} = P_x l_2 + P_y m_2 + P_z n_2.$$

Substituting Eqs. (2.18), (2.19) and (2.20) with relevant direction cosines (l_2, m_2 and n_2) into the equation above provides the following expression:

$$\sigma_{y'} = (\sigma_x l_2 + \tau_{yx} m_2 + \tau_{zx} n_2) l_2 + (\tau_{xy} l_2 + \sigma_y m_2 + \tau_{zy} n_2) m_2 + (\tau_{xz} l_2 + \tau_{yz} m_2 + \sigma_z n_2) n_2.$$

By expanding the equation above, the following is obtained:

$$\sigma_{y'} = \sigma_x l_2^2 + \tau_{yx} l_2 m_2 + \tau_{zx} l_2 n_2 + \tau_{xy} l_2 m_2 + \sigma_y m_2^2 + \tau_{zy} m_2 n_2 + \tau_{xz} l_2 n_2 + \tau_{yz} m_2 n_2$$
$$+ \sigma_z n_2^2.$$

Application of the relationships as per Eqs. (2.8), (2.9) and (2.10) in the equation above results in the following:

$$\sigma_{y'} = \sigma_x l_2^2 + \sigma_y m_2^2 + \sigma_z n_2^2 + 2\tau_{xy} l_2 m_2 + 2\tau_{yz} m_2 n_2 + 2\tau_{xz} l_2 n_2. \qquad (2.24)$$

The expression for $\sigma_{z'}$ can be derived by applying the pattern in expression (2.23) and (2.24):

$$\sigma_{z'} = \sigma_x l_3^2 + \sigma_y m_3^2 + \sigma_z n_3^2 + 2\tau_{xy} l_3 m_3 + 2\tau_{yz} m_3 n_3 + 2\tau_{xz} l_3 n_3. \qquad (2.25)$$

$\tau_{x'y'}$ can be expressed in terms of stresses (σ_{nx}, σ_{ny} and σ_{nz}) before rotation, in a form similar to Eq. (2.18). Since $\tau_{x'y'}$ is defined as the shear stress distributed along the y'-axis due to the force acting parallelly on the normal plane of x'-axis, such form needs to apply direction cosines for x' and y' with respect to x, y and z axes as per Table 2.3 (l_1, m_1, n_1, l_2, m_2 and n_2). First, applying l_2, m_2 and n_2 with Eq. (2.18) yields the following expression for $\tau_{x'y'}$:

$$\tau_{x'y'} = \sigma_{nx} l_2 + \sigma_{ny} m_2 + \sigma_{nz} n_2,$$

$$\tau_{x'y'} = P_x l_2 + P_y m_2 + P_z n_2.$$

By substituting Eqs. (2.18), (2.19) and (2.20) with the second set of direction cosines (l_1, m_1 and n_1) into the equation above, we can get the following result:

$$\tau_{x'y'} = (\sigma_x l_1 + \tau_{yx} m_1 + \tau_{zx} n_1)l_2 + (\tau_{xy} l_1 + \sigma_y m_1 + \tau_{zy} n_1)m_2 + (\tau_{xz} l_1 + \tau_{yz} m_1 + \sigma_z n_1)n_2.$$

Expanding the expression above leads to the following:

$$\tau_{x'y'} = \sigma_x l_1 l_2 + \tau_{yx} l_2 m_1 + \tau_{zx} l_2 n_1 + \tau_{xy} l_1 m_2 + \sigma_y m_1 m_2 + \tau_{zy} m_2 n_1 + \tau_{xz} l_1 n_2 + \tau_{yz} m_1 n_2$$
$$+ \sigma_z n_1 n_2$$

Application of the relationships as per Eqs. (2.8), (2.9) and (2.10) in the equation above, followed by simplification yields the following equation:

$$\tau_{x'y'} = \sigma_x l_1 l_2 + \sigma_y m_1 m_2 + \sigma_z n_1 n_2 + \tau_{xy}(l_1 m_2 + l_2 m_1) + \tau_{yz}(m_1 n_2 + m_2 n_1)$$
$$+ \tau_{xz}(l_1 n_2 + l_2 n_1). \tag{2.26}$$

The same expression is obtained if l_1, m_1 and n_1 is applied first followed by l_2, m_2 and n_2.

By referring to the pattern in expression (2.26), the expressions for $\tau_{y'z'}$ and $\tau_{x'z'}$ can be derived. For $\tau_{y'z'}$, direction cosines to be applied are l_2, m_2, n_2,l_3, m_3 and n_3, and for $\tau_{x'z'}$, direction cosines to be applied are l_1, m_1, n_1,l_3, m_3 and n_3.

$$\tau_{y'z'} = \sigma_x l_2 l_3 + \sigma_y m_2 m_3 + \sigma_z n_2 n_3 + \tau_{xy}(l_2 m_3 + l_3 m_2) + \tau_{yz}(m_2 n_3 + m_3 n_2)$$
$$+ \tau_{xz}(l_2 n_3 + l_3 n_2), \tag{2.27}$$

$$\tau_{x'z'} = \sigma_x l_1 l_3 + \sigma_y m_1 m_3 + \sigma_z n_1 n_3 + \tau_{xy}(l_1 m_3 + l_3 m_1) + \tau_{yz}(m_1 n_3 + m_3 n_1)$$
$$+ \tau_{xz}(l_1 n_3 + l_3 n_1). \tag{2.28}$$

Nine direction cosines are required to determine all stresses acting on a point after such rotation. To do so, these direction cosines (as shown in Table 2.3) are arranged in a matrix form and such a matrix is represented as a:

$$[a] = \begin{bmatrix} l_1 m_1 n_1 \\ l_2 m_2 n_2 \\ l_3 m_3 n_3 \end{bmatrix} \tag{2.29}$$

$$[a]^T = \begin{bmatrix} l_1 l_2 l_3 \\ m_1 m_2 m_3 \\ n_1 n_2 n_3 \end{bmatrix} \tag{2.30}$$

The stress components after rotation can be determined using only the components before rotation and direction cosines, with the application of the following formula:

$$[\sigma'] = [a][\sigma][a]^T$$

The matrix above can be expanded, and this leads to the following expression:

$$\begin{bmatrix} \sigma'_x & \tau_{y'x'} & \tau_{z'x'} \\ \tau_{x'y'} & \sigma'_y & \tau_{z'y'} \\ \tau_{x'z'} & \tau_{y'z'} & \sigma'_z \end{bmatrix} = \begin{bmatrix} l_1 m_1 n_1 \\ l_2 m_2 n_2 \\ l_3 m_3 n_3 \end{bmatrix} \begin{bmatrix} \sigma_x & \tau_{yx} & \tau_{zx} \\ \tau_{xy} & \sigma_y & \tau_{zy} \\ \tau_{xz} & \tau_{yz} & \sigma_z \end{bmatrix} \begin{bmatrix} l_1 l_2 l_3 \\ m_1 m_2 m_3 \\ n_1 n_2 n_3 \end{bmatrix}. \tag{2.31}$$

2.5 PRINCIPAL STRESS AND MAXIMUM SHEAR STRESS

Say, a test was conducted to determine the strength of a new material to evaluate whether it is feasible to be used as a construction material alternative. Yielding occurrs when normal stress σ_1 and shear stress τ_1 are developed in the body.

When normal stress is the concern of design, the principal stress, i.e. maximum normal stress, the material can sustain before yielding will be the objective of study. When a body yields after stress σ_a is applied to point A, the principal stress of the material will be σ_a. However, when a body yields after it is subjected to multiple stresses in different direction, e.g. σ_b and σ_c, the principal stress of the material is neither σ_b nor σ_c, but their resultant stress σ_d. For point A, there will be an infinite amount of plane with a unique inclination about three mutually orthogonal axes. The possibility that has the principal stress σ_4 acting normally to it will be known as the principal plane. On the principal plane, no shear stress exists. This concept is demonstrated in Fig. 2.15.

$$[\sigma] = \begin{bmatrix} \sigma_1 & 0 & 0 \\ 0 & \sigma_2 & 0 \\ 0 & 0 & \sigma_3 \end{bmatrix}$$

Since there is no shear stress presents, σ_1, σ_2 and σ_3 are principal stresses, σ_P. By applying the relationship in (2.21) we can get

$$\begin{bmatrix} P_x \\ P_y \\ P_z \end{bmatrix} = \begin{bmatrix} \sigma_P & 0 & 0 \\ 0 & \sigma_P & 0 \\ 0 & 0 & \sigma_P \end{bmatrix} \begin{bmatrix} l \\ m \\ n \end{bmatrix}.$$

Simplifying the expression by multiplication yields the following expression:

$$\begin{bmatrix} P_x \\ P_y \\ P_z \end{bmatrix} = \begin{bmatrix} \sigma_P l \\ \sigma_P m \\ \sigma_P n \end{bmatrix} \tag{2.32}$$

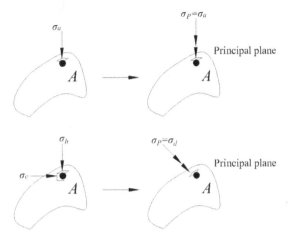

FIGURE 2.15 Principal stress on a body.

By substituting Eqs. (2.18), (2.19) and (2.20) into expression above leads to the follows:

$$\begin{bmatrix} \sigma_x l + \tau_{yx} m + \tau_{zx} n \\ \tau_{xy} l + \sigma_y m + \tau_{zy} n \\ \tau_{xz} l + \tau_{yz} m + \sigma_z n \end{bmatrix} = \begin{bmatrix} \sigma_P l \\ \sigma_P m \\ \sigma_P n \end{bmatrix}.$$

By moving all the terms to one side, the following expression is produced:

$$\begin{bmatrix} \sigma_x l + \tau_{yx} m + \tau_{zx} n \\ \tau_{xy} l + \sigma_y m + \tau_{zy} n \\ \tau_{xz} l + \tau_{yz} m + \sigma_z n \end{bmatrix} - \begin{bmatrix} \sigma_P l \\ \sigma_P m \\ \sigma_P n \end{bmatrix} = \begin{bmatrix} 0 \\ 0 \\ 0 \end{bmatrix}.$$

Simplifying the expression above gives us the following:

$$\begin{bmatrix} (\sigma_x - \sigma_P)l + \tau_{yx} m + \tau_{zx} n \\ \tau_{xy} l + (\sigma_y - \sigma_P)m + \tau_{zy} n \\ \tau_{xz} l + \tau_{yz} m + (\sigma_z - \sigma_P)n \end{bmatrix} = \begin{bmatrix} 0 \\ 0 \\ 0 \end{bmatrix}$$

Factorising the expression above with common coefficients namely l, m and n provides us the expression in the following form:

$$\begin{bmatrix} (\sigma_x - \sigma_P) & \tau_{yx} & \tau_{zx} \\ \tau_{xy} & (\sigma_y - \sigma_P) & \tau_{zy} \\ \tau_{xz} & \tau_{yz} & (\sigma_z - \sigma_P) \end{bmatrix} \begin{bmatrix} l \\ m \\ n \end{bmatrix} = \begin{bmatrix} 0 \\ 0 \\ 0 \end{bmatrix}$$

In order to obtain a potential solution to the three homogeneous linear equation given in the equation above, the determinant of the matrix above should be defined. Since the determinant only exists in a square matrix, i.e. a 3×3 matrix, we can say that

$$\begin{vmatrix} (\sigma_x - \sigma_P) & \tau_{yx} & \tau_{zx} \\ \tau_{xy} & (\sigma_y - \sigma_P) & \tau_{zy} \\ \tau_{xz} & \tau_{yz} & (\sigma_z - \sigma_P) \end{vmatrix} = 0. \tag{2.33}$$

For the 3×3 matrix below,

$$[A] = \begin{bmatrix} a_{11} & a_{12} & a_{13} \\ a_{21} & a_{22} & a_{23} \\ a_{31} & a_{32} & a_{33} \end{bmatrix}$$

the determinant can be defined as

$$Determinant, |A| = a_{11}[a_{22}a_{33} - a_{23}a_{32}] - a_{12}[a_{21}a_{33} - a_{23}a_{31}]$$
$$+ a_{13}[a_{21}a_{32} - a_{22}a_{31}].$$

Applying the formula above to find the determinant of matrix in (2.33) yields the following:

$$(\sigma_x - \sigma_P)\left[(\sigma_y - \sigma_P)(\sigma_z - \sigma_P) - (\tau_{yz})(\tau_{zy})\right] - (\tau_{yx})\left[(\tau_{xy})(\sigma_z - \sigma_P) - (\tau_{xz})(\tau_{zy})\right]$$
$$+ (\tau_{zx})\left[(\tau_{xy})(\tau_{yz}) - (\sigma_y - \sigma_P)(\tau_{xz})\right] = 0.$$

Expansion of the above leads to the following form:

$$(\sigma_x - \sigma_P)(\sigma_y - \sigma_P)(\sigma_z - \sigma_P) - (\sigma_x - \sigma_P)(\tau_{yz})(\tau_{zy}) - (\tau_{yx})(\tau_{xy})(\sigma_z - \sigma_P)$$
$$+ (\tau_{yx})(\tau_{xz})(\tau_{zy}) + (\tau_{zx})(\tau_{xy})(\tau_{yz}) - (\tau_{zx})(\sigma_y - \sigma_P)(\tau_{xz}) = 0.$$

Applying the relationships as per Eqs. (2.8), (2.9) and (2.10) in the equation above results in

$$(\sigma_x - \sigma_P)(\sigma_y - \sigma_P)(\sigma_z - \sigma_P) - (\sigma_x - \sigma_P)\tau_{yz}^2 - \tau_{xy}^2(\sigma_z - \sigma_P) + \tau_{xy}\tau_{yz}\tau_{xz} + \tau_{xy}\tau_{yz}\tau_{xz}$$
$$- \tau_{xz}^2(\sigma_y - \sigma_P) = 0.$$

Expanding the equation above yields the following:

$$\sigma_x\sigma_y\sigma_z - \sigma_x\sigma_z\sigma_P - \sigma_y\sigma_z\sigma_P + \sigma_z\sigma_P^2 - \sigma_x\sigma_y\sigma_P + \sigma_x\sigma_P^2 + \sigma_y\sigma_P^2 - \sigma_P^3 - \sigma_x\tau_{yz}^2 + \sigma_P\tau_{yz}^2$$
$$- \sigma_z\tau_{xy}^2 + \sigma_P\tau_{xy}^2 + \tau_{xy}\tau_{yz}\tau_{xz} + \tau_{xy}\tau_{yz}\tau_{xz} - \sigma_y\tau_{xz}^2 + \sigma_P\tau_{xz}^2 = 0.$$

Rearranging the equation above and factorising it with common terms produces the following equation:

$$- \sigma_P^3 + (\sigma_x + \sigma_y + \sigma_z)\sigma_P^2 - (\sigma_x\sigma_y + \sigma_y\sigma_z + \sigma_x\sigma_z - \tau_{xy}^2 - \tau_{yz}^2 - \tau_{xz}^2)\sigma_P$$
$$+ (\sigma_x\sigma_y\sigma_z - \sigma_x\tau_{yz}^2 - \sigma_y\tau_{xz}^2 - \sigma_z\tau_{xy}^2 + 2\tau_{xy}\tau_{yz}\tau_{xz}) = 0.$$

By multiplying the terms on both sides of the equation above with -1, we make the coefficient of σ_P^3 positive as follows:

$$\sigma_P^3 - (\sigma_x + \sigma_y + \sigma_z)\sigma_P^2 + \left(\sigma_x\sigma_y + \sigma_y\sigma_z + \sigma_x\sigma_z - \tau_{xy}^2 - \tau_{yz}^2 - \tau_{xz}^2\right)\sigma_P - (\sigma_x\sigma_y\sigma_z$$
$$- \sigma_x\tau_{yz}^2 - \sigma_y\tau_{xz}^2 - \sigma_z\tau_{xy}^2 + 2\tau_{xy}\tau_{yz}\tau_{xz}) = 0.$$

σ_P is the unknown that needs to be solved for. The root of the equation will be principal stresses σ_1, σ_2 and σ_3, with $\sigma_1 > \sigma_2 > \sigma_3$. Introducing new terms to simplify the coefficients of σ_P^2, σ_P and constant to the equation above yields the following:

$$\sigma_P^3 - I_1\sigma_P^2 + I_2\sigma_P - I_3 = 0 \qquad (2.34)$$

where I_1, I_2 and I_3 are defined as follows:

$$I_1 = \sigma_x + \sigma_y + \sigma_z \qquad (2.35)$$

$$I_2 = \sigma_x\sigma_y + \sigma_y\sigma_z + \sigma_x\sigma_z - \tau_{xy}^2 - \tau_{yz}^2 - \tau_{xz}^2 \qquad (2.36)$$

$$I_3 = \sigma_x\sigma_y\sigma_z - \sigma_x\tau_{yz}^2 - \sigma_y\tau_{xz}^2 - \sigma_z\tau_{xy}^2 + 2\tau_{xy}\tau_{yz}\tau_{xz} \qquad (2.37)$$

I_1, I_2 and I_3 are known as stress invariants. They are independent of the coordinate system in use. When the coordinate system coincides with directions of principal stress, normal stress components represent principal stress components, and shear stress components will be eliminated from I_1, I_2 and I_3:

$$I_1 = \sigma_1 + \sigma_2 + \sigma_3,$$

$$I_2 = \sigma_1\sigma_2 + \sigma_2\sigma_3 + \sigma_1\sigma_3,$$

$$I_3 = \sigma_1\sigma_2\sigma_3.$$

Also, Eq. (2.32) can be written as follows:

$$P_x = \sigma_1 l,$$
$$P_y = \sigma_2 m, \tag{2.38}$$
$$P_z = \sigma_3 n.$$

Substituting the relationships derived above into (2.22) leads to the following:

$$P_n = (\sigma_1 l)l + (\sigma_2 m)m + (\sigma_3 n)n,$$
$$P_n = \sigma_1 l^2 + \sigma_2 m^2 + \sigma_3 n^2. \tag{2.39}$$

By Pythagoras theorem, the resultant force of P_x, P_y and P_z, denoted by P_R, can be determined as follows:

$$P_R^2 = P_x^2 + P_y^2 + P_z^2.$$

Substituting (2.38) into the equation above produces the following:

$$P_R^2 = (\sigma_1 l)^2 + (\sigma_2 m)^2 + (\sigma_3 n)^2,$$
$$P_R^2 = \sigma_1^2 l^2 + \sigma_2^2 m^2 + \sigma_3^2 n^2. \tag{2.40}$$

P_R is also the resultant stress of normal stress, P_n and shear stress, say τ_n. The breakdown of forces is shown in Fig. 2.16.

Therefore, using the Pythagoras theorem, we can derive the following:

$$P_R^2 = P_n^2 + \tau_n^2,$$

$$\tau_n^2 = P_R^2 - P_n^2,$$

Substituting (2.39) and (2.40) into the equation above results in the following:

$$\tau_n^2 = \sigma_1^2 l^2 + \sigma_2^2 m^2 + \sigma_3^2 n^2 - (\sigma_1 l^2 + \sigma_2 m^2 + \sigma_3 n^2)^2.$$

Expanding the equation above will give us the following expression:

$$\tau_n^2 = \sigma_1^2 l^2 + \sigma_2^2 m^2 + \sigma_3^2 n^2 - \sigma_1^2 l^4 - \sigma_2^2 m^4 + \sigma_3^2 n^4 - 2\sigma_1\sigma_2 l^2 m^2 - 2\sigma_1\sigma_3 l^2 n^2$$
$$- 2\sigma_2\sigma_3 m^2 n^2.$$

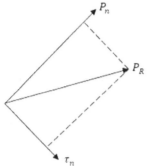

FIGURE 2.16 P_R as a resultant stress of normal and shear stresses.

Simplifying the equation above produces the following:

$$\tau_n^2 = \sigma_1^2 l^2 (1 - l^2) + \sigma_2^2 m^2 (1 - m^2) + \sigma_3^2 n^2 (1 - n^2) - 2\,\sigma_1\sigma_2 l^2 m^2 - 2\sigma_1\sigma_3 l^2 n^2$$
$$- 2\,\sigma_2\sigma_3 m^2 n^2.$$

(2.41)

From Eq. (2.13), the following equations can be derived:

$$1 - l^2 = m^2 + n^2,$$

$$1 - m^2 = l^2 + n^2,$$

$$1 - n^2 = l^2 + m^2.$$

Substituting the relationships above into (2.41) yields the following:

$$\tau_n^2 = \sigma_1^2 l^2 (m^2 + n^2) + \sigma_2^2 m^2 (l^2 + n^2) + \sigma_3^2 n^2 (l^2 + m^2) - 2\,\sigma_1\sigma_2 l^2 m^2 - 2\sigma_1\sigma_3 l^2 n^2$$
$$- 2\,\sigma_2\sigma_3 m^2 n^2.$$

Expanding the equation above produces the following:

$$\tau_n^2 = \sigma_1^2 l^2 m^2 + \sigma_1^2 l^2 n^2 + \sigma_2^2 l^2 m^2 + \sigma_2^2 m^2 n^2 + \sigma_3^2 l^2 n^2 + \sigma_3^2 m^2 n^2 - 2\,\sigma_1\sigma_2 l^2 m^2$$
$$- 2\sigma_1\sigma_3 l^2 n^2 - 2\,\sigma_2\sigma_3 m^2 n^2.$$

Simplifying the equation above by factorizing with common terms leads to the following:

$$\tau_n^2 = l^2 m^2 (\sigma_1^2 + \sigma_2^2 - 2\,\sigma_1\sigma_2) + l^2 n^2 (\sigma_1^2 + \sigma_3^2 - 2\sigma_1\sigma_3) + m^2 n^2 (\sigma_2^2 + \sigma_3^2 - 2\,\sigma_2\sigma_3).$$

Simplifying the equation above yields the following expression:

$$\tau_n^2 = l^2 m^2 (\sigma_1 - \sigma_2)^2 + l^2 n^2 (\sigma_1 - \sigma_3)^2 + m^2 n^2 (\sigma_2 - \sigma_3)^2. \tag{2.42}$$

The principal stress is maximum shear stress when the planes are angled 45°. Fig. 2.17 shows the conditions for maximum shear stress to occur. For example, when the axes are rotated about axes 1, there will be no shear stress in the 1-direction, and the shear stress acting in the other two (2 and 3) directions is maximum:

$$\frac{\partial \tau_n}{\partial l} = 0 \rightarrow l = 0, \ m = n = \ \cos 45° = \pm \frac{1}{\sqrt{2}},$$

$$\frac{\partial \tau_n}{\partial m} = 0 \rightarrow m = 0, \ l = n = \ \cos 45° = \pm \frac{1}{\sqrt{2}}, \tag{2.43}$$

$$\frac{\partial \tau_n}{\partial n} = 0 \rightarrow n = 0, \ l = m = \ \cos 45° = \pm \frac{1}{\sqrt{2}}.$$

Substituting the values of m and n obtained for case $\frac{\partial \tau_n}{\partial l} = 0$ as in (2.43) into (2.42) yields the following:

$$\tau_{2,3,\ max}^2 = \left(\pm \frac{1}{\sqrt{2}} \right)^2 \left(\pm \frac{1}{\sqrt{2}} \right)^2 (\sigma_2 - \sigma_3)^2,$$

$$\tau_{2,3,\ max}^2 = \frac{(\sigma_2 - \sigma_3)^2}{4},$$

$$\tau_{2,3,max} = \pm \frac{(\sigma_2 - \sigma_3)}{2}.$$

Shear stress components for other planes can be expressed in a similar way, as follows:

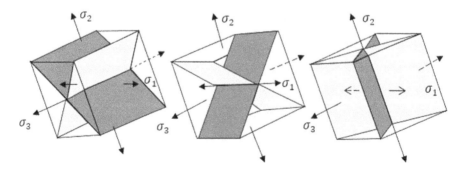

FIGURE 2.17 Planes of maximum shear stresses.

$$\tau_{1,3,\,max} = \pm\frac{(\sigma_1 - \sigma_3)}{2}$$
$$\tau_{1,2,max} = \pm\frac{(\sigma_1 - \sigma_2)}{2}$$

(2.44)

Corresponding to the maximum shear stresses, the normal stresses are defined as follow:

$$\sigma_{n1} = \pm\frac{\sigma_1 + \sigma_2}{2}$$

$$\sigma_{n2} = \pm\frac{\sigma_2 + \sigma_3}{2}$$

$$\sigma_{n3} = \pm\frac{\sigma_1 + \sigma_3}{2}$$

2.6   DEVIATORIC STRESS

When stresses are developed in non-rigid body, dilation and distortion occur. In relation to that, the developed stresses are divided into two components for the convenience of analysis: hydrostatic stress, which is responsible for dilation and deviatoric stress, which is responsible for distortion.

Hydrostatic stress is the average of three principal stresses along the respective axes and acting along all three axes. Deviatoric stress is the difference between principal stress and hydrostatic stress along all three axes. Consider a body develops different amount of stresses along three mutually orthogonal axes. Different strains, which are hydrostatic strain and deviatoric strain corresponding to their counterparts, are shown. It can be observed that when deviatoric stress along an axis is high, the distortion of body along that axis is significant, as indicated in Fig. 2.18.

The hydrostatic stress can be written as follows:

$$\sigma_m = \frac{\sigma_x + \sigma_y + \sigma_z}{3}.$$

(2.45)

Applying the relationship in Eq. (2.35) to the equation above yields the following expression:

$$\sigma_m = \frac{I_1}{3}.$$

(2.46)

Dilation due to hydrostatic stress is the uniform displacement of particles across a solid, whether it is towards or away from each other. In contrast, distortion due to deviatoric stress is differential displacement of particles across the solid. Therefore, distortion in this case is very likely to break the bond between particles compared to

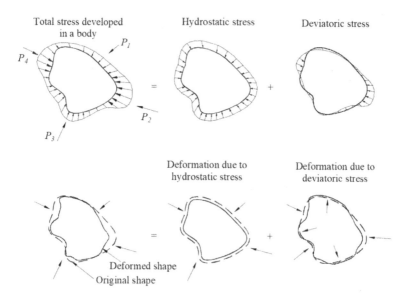

FIGURE 2.18   Hydrostatic and deviatoric stresses.

dilation. In other words, deviatoric stress is the kind of stress responsible for yielding. It is essential for engineers and material scientists when the plastic behaviour of solid is concerned.

Let the total stress be σ. The deviatoric stress, σ', is defined as follows

$$\sigma' = \sigma - \sigma_m.$$

Substituting (2.46) into the equation above gives the following equation:

$$\sigma' = \sigma - \frac{I_1}{3}.$$

In other words, we can express the total stress in the following form:

$$\sigma = \sigma' + \frac{I_1}{3}.$$

Substituting the equation above into Eq. (2.34) we can get

$$\left(\sigma' + \frac{I_1}{3}\right)^3 - I_1\left(\sigma' + \frac{I_1}{3}\right)^2 + I_2\left(\sigma' + \frac{I_1}{3}\right) - I_3 = 0.$$

Expanding the equation above leads to the following:

$$\sigma'^3 + \frac{2I_1\sigma'^2}{3} + \frac{I_1^2\sigma'}{9} + \frac{I_1\sigma'^2}{3} + \frac{2I_1^2\sigma'}{9} + \frac{I_1^3}{27} - I_1\sigma'^2 - \frac{2I_1^2\sigma'}{3} - \frac{I_1^3}{9} + I_2\sigma' + \frac{I_1I_2}{3}$$
$$- I_3 = 0$$

Rearranging the equation above and factorising it with common terms yields the following:

$$\sigma'^3 + \left(\frac{2I_1}{3} + \frac{I_1}{3} - I_1\right)\sigma'^2 + \left(\frac{I_1^2}{9} + \frac{2I_1^2}{9} - \frac{2I_1^2}{3} + I_2\right)\sigma' - \frac{I_1^3}{9} + \frac{I_1^3}{27} + \frac{I_1I_2}{3} - I_3$$
$$= 0.$$

Simplification of the equation above produces the following expression:

$$\sigma'^3 - \left(\frac{I_1^2}{3} - I_2\right)\sigma' - \left(\frac{2I_1^3}{27} - \frac{I_1I_2}{3} + I_3\right) = 0.$$

σ' is the unknown that needs to be solved. Introducing new terms to simplify the coefficients of σ' and constant to the equation above leads to the following:

$$\sigma'^3 - J_2\sigma' - J_3 = 0$$

where J_2 and J_3 are known as invariants of deviatoric stresses defined as follows:

$$J_2 = \frac{I_1^2}{3} - I_2. \tag{2.47}$$

$$J_3 = \frac{2I_1^3}{27} - \frac{I_1I_2}{3} + I_3. \tag{2.48}$$

Similar to Eq. (2.35), J_1 is defined as:

$$J_1 = \sigma'_x + \sigma'_y + \sigma'_z.$$

Substituting (2.46) into (2.48) yields the following equation:

$$J_3 = I_3 - I_2\sigma_m + 2\sigma_m^3.$$

2.7 OCTAHEDRAL STRESS

An octahedral plane inclines equally against three mutually orthogonal axes. It is possible to form eight octahedral planes and an octahedron as a result, as shown in Fig. 2.19.

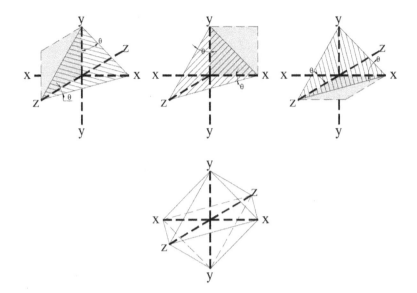

FIGURE 2.19 Octahedral plane and octahedron along three mutually orthogonal axes.

Consider an octahedral plane *ABC* below. As an octahedral plane, the lengths *OA*, *OB* and *OC* are equal, as shown in Fig. 2.20.

Applying the relationship in Eq. (2.12) into Eq. (2.13) yields the following

$$1 = \left(\frac{x}{r}\right)^2 + \left(\frac{y}{r}\right)^2 + \left(\frac{z}{r}\right)^2$$

Simplification of the equation to express *r* in terms of the orthogonal axes gives us the following expression:

$$r^2 = x^2 + y^2 + z^2.$$

x, *y* and *z* represent *OC*, *OB* and *OA*, respectively, and *OA* = *OB* = *OC*, therefore, the above equation can be written as:

$$r^2 = OA^2 + OA^2 + OA^2.$$

Simplifying the equation above results in the following:

$$r^2 = 3OA^2$$
$$r = \pm\sqrt{3}\,OA \tag{2.49}$$

By applying the definition of *r* as per Eq. (2.49) and *x* = *OA* into Eq. (2.12), we can get:

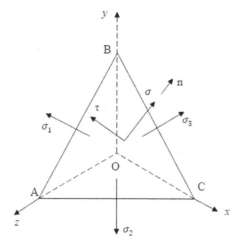

FIGURE 2.20 Stresses acting on an octahedral plane.

$$l = \frac{OA}{\pm\sqrt{3}\,OA}$$
$$l = \pm\frac{1}{\sqrt{3}} \tag{2.50}$$

Similarly, we can write the following equations for other direction cosine components:

$$m = \pm\frac{1}{\sqrt{3}}$$
$$n = \pm\frac{1}{\sqrt{3}} \tag{2.51}$$

Stresses acting on an octahedron around point O will have the same intensity of normal and shear stresses at that point. These stresses are known as octahedral normal stress, σ_{oct} and octahedral shear stress, τ_{oct}, as indicated in Fig. 2.21.

By applying (2.39), P_n will be interchangeable with σ_{oct} if the plane is an octahedral plane, and this leads to the following:

$$\sigma_{oct} = \sigma_1 l^2 + \sigma_2 m^2 + \sigma_3 n^2$$

The following is obtained by applying relationships in Eqs. (2.50) and (2.51) into equation:

$$\sigma_{oct} = \sigma_1\left(\pm\frac{1}{\sqrt{3}}\right)^2 + \sigma_2\left(\pm\frac{1}{\sqrt{3}}\right)^2 + \sigma_3\left(\pm\frac{1}{\sqrt{3}}\right)^2$$

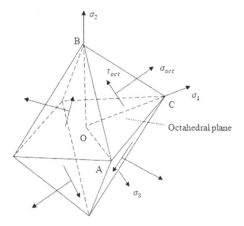

FIGURE 2.21 Stresses acting on an octahedron.

Simplifying the equation above leads to the following equation:

$$\sigma_{oct} = \frac{\sigma_1 + \sigma_2 + \sigma_3}{3} \tag{2.52}$$

By substituting Eq. (2.35) into the equation above, we can obtain

$$\sigma_{oct} = \frac{I_1}{3}.$$

Applying the relationships in Eqs. (2.50) and (2.51) into (2.40), and defining P_R as $\sigma_{R,oct}$, which denotes resultant stress of octahedral normal and shear stresses, yields the follows:

$$\sigma_{R,oct}^2 = \sigma_1^2 \left(\pm \frac{1}{\sqrt{3}} \right)^2 + \sigma_2^2 \left(\pm \frac{1}{\sqrt{3}} \right)^2 + \sigma_3^2 \left(\pm \frac{1}{\sqrt{3}} \right)^2$$

The following equation is obtained through simplification of the equation above:

$$\sigma_{R,oct}^2 = \frac{1}{3} (\sigma_1^2 + \sigma_2^2 + \sigma_3^2). \tag{2.53}$$

Using the Pythagoras theorem, we can write the following expression

$$\tau_{oct}^2 = \sigma_{R,oct}^2 - \sigma_{oct}^2.$$

By substituting Eqs. (2.52) and (2.53) into equation above produces the following:

$$\tau_{oct}^2 = \frac{1}{3}(\sigma_1^2 + \sigma_2^2 + \sigma_3^2) - \left(\frac{\sigma_1 + \sigma_2 + \sigma_3}{3}\right)^2.$$

Expanding the equation above yields the following equation

$$\tau_{oct}^2 = \left(\frac{\sigma_1^2}{3} + \frac{\sigma_2^2}{3} + \frac{\sigma_3^2}{3}\right) - \left(\frac{\sigma_1^2 + \sigma_2^2 + \sigma_3^2 + 2\,\sigma_1\sigma_2 + 2\,\sigma_2\sigma_3 + 2\,\sigma_1\,\sigma_3}{9}\right).$$

Simplify the equation above and we get:

$$\tau_{oct}^2 = \frac{\sigma_1^2}{3} + \frac{\sigma_2^2}{3} + \frac{\sigma_3^2}{3} - \frac{\sigma_1^2}{9} - \frac{\sigma_2^2}{9} - \frac{\sigma_3^2}{9} - \frac{2\,\sigma_1\sigma_2}{9} - \frac{2\sigma_2\sigma_3}{9} - \frac{2\sigma_1\sigma_3}{9},$$

$$\tau_{oct}^2 = \frac{2\,\sigma_1^2}{9} + \frac{2\,\sigma_2^2}{9} + \frac{2\,\sigma_3^2}{9} - \frac{2\,\sigma_1\sigma_2}{9} - \frac{2\,\sigma_2\sigma_3}{9} - \frac{2\,\sigma_1\,\sigma_3}{9}.$$

Factorisation of the equation above produces the following

$$\tau_{oct}^2 = \frac{(\sigma_1 - \sigma_2)^2 + (\sigma_2 - \sigma_3)^2 + (\sigma_3 - \sigma_1)^2}{9}.$$

Taking the square root on both sides of the equation leads to the following

$$\tau_{oct} = \pm\frac{1}{3}\sqrt{(\sigma_1 - \sigma_2)^2 + (\sigma_2 - \sigma_3)^2 + (\sigma_3 - \sigma_1)^2} \qquad (2.54)$$

2.8 PLANE STRESS

Consider the thickness, dz, of a body (say, along the z-axis) is much smaller than its length, dy and width, dx (along x and y axes). When force is exerted in a direction normal to the z-axis, stresses are developed only on the xy-plane. By inspecting the sectional planes normal to the z-axis, it can be observed that the thickness of body is too small for the stress to vary along the z-direction. Since thickness does not play any role in stress distribution within the body, this scenario can be simplified into plane stress scenario, as demonstrated in Fig. 2.22.

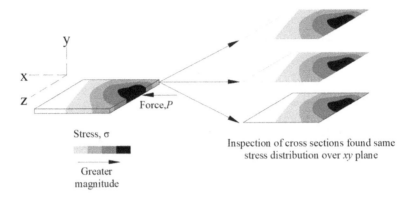

FIGURE 2.22 Plane stress.

Under the plane stress scenario, the component of stresses as per Eq. (2.3) can be simplified as follows:

$$\sigma = \begin{bmatrix} \sigma_X & \tau_{xy} & 0 \\ \tau_{yx} & \sigma_Y & 0 \\ 0 & 0 & 0 \end{bmatrix}.$$

3 Strain

3.1 DEFORMATION AND STRAIN

The overall displacement of the particles in such body is known as deformation. By deforming, the body converts work done to elastic strain energy and kinetic energy. If there is no more external force exerted on it, the body will stop deform after such energy conversion is completed. When external forces reduced, the elastic strain energy will exert elastic restoring force, which will have its work done converted to kinetic energy and moves the particles back to their original position, if the bond between particles has not been break. There are two types of deformation: rigid body motion and non-rigid body deformation.

Rigid body motion occurs when the exerted force is enough to change the motion state of a body but insufficient stress is developed to overcome its attraction force. Rigid body motion includes translation and rotation. In translation, all points on the body have displacement of same magnitude and in the same direction. In rotation, all points on the body have angular displacement of same angle and in the same direction, except the points that lie along the axis of rotation.

Non-rigid body deformation occurs when stress developed is enough to overcome the attraction force, regardless the ability of force to change a body's motion state. This type of deformation includes distortion and dilation, which is in fact the result of differential translation and rotation. Distortion is a process where a body change its shape, while dilation is a process where a body change its volume. Fig. 3.1 shows some examples of rigid and non-rigid body deformation in a girder.

As an observable and measurable quantity, deformation is an aspect that material scientists look up to identify the material's mechanical properties. However, direct measurement of a body's deformation is not reliable: a very short rubber stick deforms less than a very long steel rod. One can wrongfully conclude that rubber is more rigid

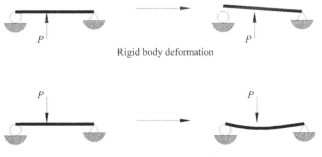

Rigid body deformation

Non-rigid body deformation

FIGURE 3.1 Examples of rigid body and non-rigid body deformations in girder.

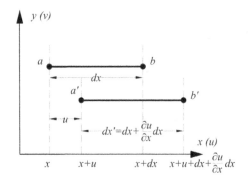

FIGURE 3.2 Normal strain in a 1-D element.

than steel if he interpreted test result this way. Also, when a solid is experiencing rigid body deformation, no differential deformation takes place throughout the body. Differential deformation only takes place during non-rigid body deformation, which is an indication of development of stress inside within the solid. For these reasons, strain, i.e a way to determine the severity of deformation using ratio of solid's deformation to its original length along a specific direction, is looked up.

Assuming there is a 1-D element subjected to normal force and exhibited both rigid and non-rigid body deformation: translation and elongation (distortion). Let the distance between two nodes of the element, namely a and b be dx. Also, let the deformation along x and y axes be u and v, respectively. The correlations between the defined parameters are as shown in Fig. 3.2.

Translation caused the element to move u unit along positive x direction, and elongation added on extra deformation along x direction. This additional deformation is proportional to the length of element, which can be expressed as $\frac{\partial u}{\partial x}$. For element with length of dx, the total deformation caused by elongation is $\frac{\partial u}{\partial x} \times dx$.

The equation of normal strain is given as follows:

$$\epsilon = \frac{\Delta L}{L} \tag{3.1}$$

where ϵ is normal strain, ΔL is the change in longitudinal length and L is the original length of body.

The following equation yielded by applying the case as illustrated in Fig. 3.2 into Eq. (3.1):

$$
\begin{aligned}
\epsilon_x &= \frac{a'b' - ab}{ab} \\
&= \frac{dx + \frac{\partial u}{\partial x}dx - dx}{dx} \tag{3.2} \\
&= \frac{\partial u}{\partial x}
\end{aligned}
$$

This equation proves that rigid body deformation, i.e translation, does not influence the strain of body. Similarly, the strain expression for another two axes can be written in terms of v and w, which is deformation along z direction as the following:

$$\epsilon_y = \frac{\partial v}{\partial y} \tag{3.3}$$

$$\epsilon_z = \frac{\partial w}{\partial z} \tag{3.4}$$

Assuming a 2-D element subjected to shear force and exhibited both rigid and non-rigid body deformation: translation and distortion. Let the nodes of the element be A, B, C and D, and the dimension of such element along x and y axes be dx and dy respectively. Similarly, deformation along x and y axes will be u and v respectively, as shown in Fig. 3.3.

Translation caused the element to move u unit along positive x direction and v unit along positive y direction. Distortion is caused by the rotation of edge AB and AD, which now becomes $A'B'$ and $A'D'$. Shear strain is defined as the total angular deformation of a body. In this case, it is defined as below:

$$\gamma_{xy} = \tan\beta_1 + \tan\beta_2 \tag{3.5}$$

If infinitesimal deformations are considered, then by small angle approximation,

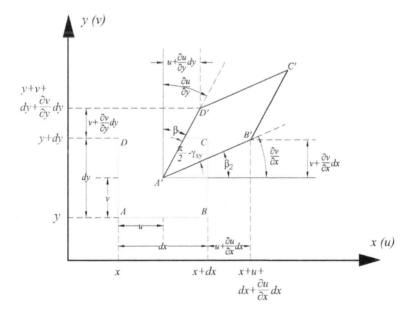

FIGURE 3.3   Shear strain in a 2-D element.

$$\tan \theta \approx \theta, \quad \text{where } \theta \approx 0 \text{ and in unit radian}$$

Therefore, Eq. (3.5) can be expressed as follows:

$$\gamma_{xy} \approx \beta_1 + \beta_2$$
$$\gamma_{xy} \approx \frac{\partial u}{\partial y} + \frac{\partial v}{\partial x} \tag{3.6}$$

Similarly, we can write the following shear strain expressions for another two axes:

$$\gamma_{xz} \approx \frac{\partial u}{\partial z} + \frac{\partial w}{\partial x} \tag{3.7}$$

$$\gamma_{yz} \approx \frac{\partial v}{\partial z} + \frac{\partial w}{\partial x} \tag{3.8}$$

By combining all the aforementioned cases of deformation, total deformation along each axis is defined as below:

$$
\begin{aligned}
du &= \frac{\partial u}{\partial x}dx + \frac{\partial u}{\partial y}dy + \frac{\partial u}{\partial z}dz \\
dv &= \frac{\partial v}{\partial x}dx + \frac{\partial v}{\partial y}dy + \frac{\partial v}{\partial z}dz \\
dw &= \frac{\partial w}{\partial x}dx + \frac{\partial w}{\partial y}dy + \frac{\partial w}{\partial z}dz
\end{aligned} \tag{3.9}
$$

In matrix form, the equations above are expressed as follows:

$$
\begin{bmatrix} du \\ dv \\ dz \end{bmatrix} =
\begin{bmatrix}
\frac{\partial u}{\partial x} & \frac{\partial u}{\partial y} & \frac{\partial u}{\partial z} \\
\frac{\partial v}{\partial x} & \frac{\partial v}{\partial y} & \frac{\partial v}{\partial z} \\
\frac{\partial w}{\partial x} & \frac{\partial w}{\partial y} & \frac{\partial w}{\partial z}
\end{bmatrix}
\begin{bmatrix} dx \\ dy \\ dz \end{bmatrix} \tag{3.10}
$$

The relationship in Eq. (3.1) can be rewrite as this:

$$\Delta L = \epsilon L$$

The following is obtained when comparison of equation above with Eq. (3.10) is made:

$$
[\epsilon] =
\begin{bmatrix}
\frac{\partial u}{\partial x} & \frac{\partial u}{\partial y} & \frac{\partial u}{\partial z} \\
\frac{\partial v}{\partial x} & \frac{\partial v}{\partial y} & \frac{\partial v}{\partial z} \\
\frac{\partial w}{\partial x} & \frac{\partial w}{\partial y} & \frac{\partial w}{\partial z}
\end{bmatrix} \tag{3.11}
$$

Shear strain is the product of coupled shear stress.

In Fig. 3.3, β_1 and β_2 are the effect of coupled shear stress on the body. Since the applied stresses are the same, and the variation of material properties over infinitesimal element is negligible, it can be concluded that:

$$\beta_1 = \beta_2$$

Therefore, from Eq. (3.6), shear strain is defined as below:

$$\frac{1}{2}\gamma_{xy} = \frac{\partial u}{\partial y} = \frac{\partial v}{\partial x} \tag{3.12}$$

For other two shear strain components we can write their expressions in the similar fashion, as shown below:

$$\frac{1}{2}\gamma_{yz} = \frac{\partial v}{\partial z} = \frac{\partial w}{\partial y}$$
$$\frac{1}{2}\gamma_{xz} = \frac{\partial u}{\partial z} = \frac{\partial w}{\partial x} \tag{3.13}$$

By substituting relationships in Eqs. (3.2), (3.3), (3.4), (3.12) and (3.13) into Eq. (3.11) yields the following:

$$[\epsilon] = \begin{bmatrix} \epsilon_x & \frac{1}{2}\gamma_{yx} & \frac{1}{2}\gamma_{zx} \\ \frac{1}{2}\gamma_{xy} & \epsilon_y & \frac{1}{2}\gamma_{zy} \\ \frac{1}{2}\gamma_{xz} & \frac{1}{2}\gamma_{yz} & \epsilon_z \end{bmatrix} \tag{3.14}$$

3.2   LAGRANGIAN DESCRIPTION

Since deformation is a result of change in velocity of points on a body, it can be expressed as a function of time. Take an infinitesimal element on a solid as shown in Fig. 3.4 as an example. When external force is exerted, the point A and B of this element moves with different velocity. At the time t_1, point A reaches point A'.

Lagrangian description is used to express the function of deformation based on initial condition and time. In solid mechanics, initial conditions, e.g. geometry and boundary condition, are usually specified or easy to define even if they are not and thus, Lagrangian description is a kind of expression that eases the process to determine deformation of any point. Therefore, the vector of point A' (after deformation) is interested and it can be written as:

$$a' = a'(x, \ y, \ z, \ t)$$

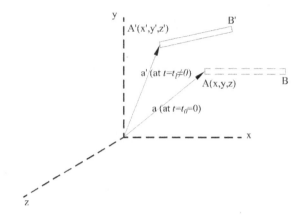

FIGURE 3.4 Deformation of an infinitesimal element.

Since position vector a is consist of position component in x, y and z axes, namely x, y and z, the function can also be written as:

$$x' = x'(x, \ y, \ z, \ t)$$

$$y' = y'(x, \ y, \ z, \ t)$$

$$z' = z'(x, \ y, \ z, \ t)$$

Let U be the vector of displacement attained by a point upon exertion of force at time t_1, and consists of displacement components along x, y and z axes, namely u, v and w. This vector is equals to the change in position vector from initial condition to time t_1:

$$[U] = [a'] - [a]$$

The following is obtained after rewriting the displacement vector in terms of displacement components in x, y and z axes:

$$u = x' - x$$

$$v = y' - y$$

$$w = z' - z$$

By expressing the coordinates of points A, B, A' and B' in terms of position components (x, y, z) and displacement components (u, v, w) results as follows:

Point	x-coordinate	y-coordinate	z-coordinate
A	x	y	z
B	$x + dx$	$y + dy$	$z + dz$
A'	x' or $x + u$	y' or $y + v$	z' or $z + w$
B'	$x' + dx'$ or $x + u + dx + du$	$y' + dy'$ or $y + v + dy + dv$	$z' + dz'$ or $z + w + dz + dw$

The original length of element, which is also the distance between A and B, say dS can be expressed through Pythagoras theorem:

$$dS^2 = AB^2$$

$$dS^2 = [(x + dx) - x]^2 + [(y + dy) - y]^2 + [(z + dz) - z]^2$$

$$dS^2 = (dx)^2 + (dy)^2 + (dz)^2$$

Using the similar fashion, the length of element after deformation is defined as:

$$dS'^2 = (dx')^2 + (dy')^2 + (dz')^2$$

$$dS'^2 = (dx + du)^2 + (dy + dv)^2 + (dz + dw)^2$$

The difference between the length before and after deformation is the amount of deformation itself. If such difference is zero, then the solid experienced rigid body deformation. By finding such difference we can get the following:

$$(dS')^2 - (dS)^2 = [(dx + du)^2 + (dy + dv)^2 + (dz + dw)^2] - (dx^2 + dy^2 + dz^2)$$

Let dL be the amount of deformation, then expand the terms yields the follows:

$$dL^2 = [(dx)^2 + 2du\,dx + (du)^2 + (dy)^2 + 2\,dv\,dy + (dv)^2 + (dz)^2 + 2\,dw\,dz + dw^2 \\ - (dx^2 + dy^2 + dz^2)]$$

The following is resulted after simplification:

$$dL^2 = 2du\,dx + du^2 + 2\,dv\,dy + dv^2 + 2\,dw\,dz + dw^2$$

By substituting the expression of du, dv and dw from Eq. (3.9) into equation above leads to the following expression:

$$dL^2 = 2dx\left[\frac{\partial u}{\partial x}dx + \frac{\partial u}{\partial y}dy + \frac{\partial u}{\partial z}dz\right] + \left[\frac{\partial u}{\partial x}dx + \frac{\partial u}{\partial y}dy + \frac{\partial u}{\partial z}dz\right]^2$$

$$+ 2dy\left[\frac{\partial v}{\partial x}dx + \frac{\partial v}{\partial y}dy + \frac{\partial v}{\partial z}dz\right] + \left[\frac{\partial v}{\partial x}dx + \frac{\partial v}{\partial y}dy + \frac{\partial v}{\partial z}dz\right]^2$$

$$+ 2dz\left[\frac{\partial w}{\partial x}dx + \frac{\partial w}{\partial y}dy + \frac{\partial w}{\partial z}dz\right] + \left[\frac{\partial w}{\partial x}dx + \frac{\partial w}{\partial y}dy + \frac{\partial w}{\partial z}dz\right]^2$$

Expansion of equation above produced the following form:
Expansion of equation above produced the following form:

$$dL^2 = 2dx^2\left(\frac{\partial u}{\partial x}\right) + 2dxdy\left(\frac{\partial u}{\partial y}\right) + 2dxdz\left(\frac{\partial u}{\partial z}\right) + \left(\frac{\partial u}{\partial x}\right)^2 dx^2 + \left(\frac{\partial u}{\partial y}\right)^2 dy^2 + \left(\frac{\partial u}{\partial z}\right)^2 dz^2$$

$$+ 2dxdy\left(\frac{\partial u}{\partial x}\right)\left(\frac{\partial u}{\partial y}\right) + 2dydz\left(\frac{\partial u}{\partial y}\right)\left(\frac{\partial u}{\partial z}\right) + 2dxdz\left(\frac{\partial u}{\partial x}\right)\left(\frac{\partial u}{\partial z}\right) + 2dxdy\left(\frac{\partial v}{\partial x}\right)$$

$$+ 2dy^2\left(\frac{\partial v}{\partial y}\right) + 2dydz\left(\frac{\partial v}{\partial z}\right) + \left(\frac{\partial v}{\partial x}\right)^2 (dx)^2 + \left(\frac{\partial v}{\partial y}\right)^2 (dy)^2 + \left(\frac{\partial v}{\partial z}\right)^2 (dz)^2$$

$$+ 2dxdy\left(\frac{\partial v}{\partial x}\right)\left(\frac{\partial v}{\partial y}\right) + 2dydz\left(\frac{\partial w}{\partial y}\right)\left(\frac{\partial v}{\partial z}\right) + 2dxdz\left(\frac{\partial v}{\partial x}\right)\left(\frac{\partial v}{\partial z}\right) + dxdz\left(\frac{\partial w}{\partial x}\right)$$

$$+ 2dydz\left(\frac{\partial w}{\partial y}\right) + 2dz^2\left(\frac{\partial w}{\partial z}\right) + dx^2\left(\frac{\partial w}{\partial x}\right)^2 + dy^2\left(\frac{\partial w}{\partial y}\right)^2 + dz^2\left(\frac{\partial w}{\partial z}\right)^2$$

$$+ 2dxdy\left(\frac{\partial w}{\partial x}\right)\left(\frac{\partial w}{\partial y}\right) + 2dydz\left(\frac{\partial w}{\partial y}\right)\left(\frac{\partial w}{\partial z}\right) + 2dxdz\left(\frac{\partial w}{\partial x}\right)\left(\frac{\partial w}{\partial z}\right) \tag{3.15}$$

Theoretically, any amount of force can deform a body by a certain degree, and that change in geometry will then affect the further response of body towards existing force or force to be exerted. Geometric non-linearity relationships are introduced to tackle the change in body's behaviour throughout the loading period. Such relationships are expressed as below:

$$L_{xx} = \frac{\partial u}{\partial x} + \frac{1}{2}\left[\left(\frac{\partial u}{\partial x}\right)^2 + \left(\frac{\partial v}{\partial x}\right)^2 + \left(\frac{\partial w}{\partial z}\right)^2\right]$$

$$L_{yy} = \left(\frac{\partial v}{\partial y}\right) + \frac{1}{2}\left[\left(\frac{\partial u}{\partial y}\right)^2 + \left(\frac{\partial v}{\partial y}\right)^2 + \left(\frac{\partial w}{\partial y}\right)^2\right]$$

$$L_{zz} = \left(\frac{\partial w}{\partial z}\right) + \frac{1}{2}\left[\left(\frac{\partial u}{\partial z}\right)^2 + \left(\frac{\partial v}{\partial x}\right)^2 + \left(\frac{\partial w}{\partial z}\right)^2\right]$$

$$L_{xy} = \left(\frac{\partial u}{\partial y}\right) + \left(\frac{\partial v}{\partial x}\right) + \left(\frac{\partial u}{\partial x}\right)\left(\frac{\partial u}{\partial y}\right) + \left(\frac{\partial v}{\partial x}\right)\left(\frac{\partial v}{\partial y}\right) + \left(\frac{\partial w}{\partial x}\right)\left(\frac{\partial w}{\partial y}\right)$$

$$L_{yz} = \left(\frac{\partial v}{\partial z}\right) + \left(\frac{\partial w}{\partial y}\right) + \left(\frac{\partial u}{\partial y}\right)\left(\frac{\partial u}{\partial z}\right) + \left(\frac{\partial v}{\partial y}\right)\left(\frac{\partial v}{\partial z}\right) + \left(\frac{\partial w}{\partial}\right)\left(\frac{\partial w}{\partial}\right)$$

$$L_{xz} = \left(\frac{\partial u}{\partial z}\right) + \left(\frac{\partial w}{\partial x}\right) + \left(\frac{\partial u}{\partial x}\right)\left(\frac{\partial u}{\partial z}\right) + \left(\frac{\partial v}{\partial x}\right)\left(\frac{\partial v}{\partial z}\right) + \left(\frac{\partial w}{\partial x}\right)\left(\frac{\partial w}{\partial z}\right)$$

$$\tag{3.16}$$

The following is obtained with rearrangement and simplification of Eq. (3.15) using the relationships above:

$$
\begin{aligned}
dL^2 = \quad & 2dx^2\left[\left(\frac{\partial u}{\partial x}\right) + \frac{1}{2}\left[\left(\frac{\partial u}{\partial x}\right)^2 + \left(\frac{\partial v}{\partial x}\right)^2 + \left(\frac{\partial w}{\partial z}\right)^2\right]\right] \\
& + 2dy^2\left[\left(\frac{\partial v}{\partial y}\right) + \frac{1}{2}\left[\left(\frac{\partial u}{\partial y}\right)^2 + \left(\frac{\partial v}{\partial y}\right)^2 + \left(\frac{\partial w}{\partial y}\right)^2\right]\right] \\
& + 2dz^2\left[\left(\frac{\partial w}{\partial z}\right) + \frac{1}{2}\left[\left(\frac{\partial u}{\partial z}\right)^2 + \left(\frac{\partial v}{\partial x}\right)^2 + \left(\frac{\partial w}{\partial z}\right)^2\right]\right] \\
& + 2dx\,dy\left[\left(\frac{\partial u}{\partial y}\right) + \left(\frac{\partial v}{\partial x}\right) + \left(\frac{\partial u}{\partial x}\right)\left(\frac{\partial u}{\partial y}\right) + \left(\frac{\partial v}{\partial x}\right)\left(\frac{\partial v}{\partial y}\right) + \left(\frac{\partial w}{\partial x}\right)\left(\frac{\partial w}{\partial y}\right)\right] \\
& + 2dy\,dz\left[\left(\frac{\partial v}{\partial z}\right) + \left(\frac{\partial w}{\partial y}\right) + \left(\frac{\partial u}{\partial y}\right)\left(\frac{\partial u}{\partial z}\right) + \left(\frac{\partial v}{\partial y}\right)\left(\frac{\partial v}{\partial z}\right) + \left(\frac{\partial w}{\partial}\right)\left(\frac{\partial w}{\partial}\right)\right] \\
& + 2dx\,dz\left[\left(\frac{\partial u}{\partial z}\right) + \left(\frac{\partial w}{\partial x}\right) + \left(\frac{\partial u}{\partial x}\right)\left(\frac{\partial u}{\partial z}\right) + \left(\frac{\partial v}{\partial x}\right)\left(\frac{\partial v}{\partial z}\right) + \left(\frac{\partial w}{\partial x}\right)\left(\frac{\partial w}{\partial z}\right)\right]
\end{aligned}
$$

$$
dL^2 = 2\left[L_{xx}dx^2 + L_{yy}dy^2 + L_{zz}dz^2 + L_{xy}\,dx\,dy + L_{yz}\,dy\,dz + L_{xz}\,dx\,dz\right] \quad (3.17)
$$

When the strains, i.e $\frac{\partial u}{\partial x}, \frac{\partial u}{\partial y}, \frac{\partial u}{\partial z}, \frac{\partial v}{\partial x}, \frac{\partial v}{\partial y}, \frac{\partial v}{\partial z}, \frac{\partial w}{\partial x}, \frac{\partial w}{\partial y}$ and $\frac{\partial w}{\partial z}$, are infinitesimal, their product with each other, including second order of each term will be too insignificant to consider. Therefore, by neglecting the second order terms in Eq. (3.16) the non-linearity relationship will be transformed into linearity relationship as below:

$$
\begin{aligned}
L_{xx} &= \frac{\partial u}{\partial x} = \varepsilon_x \\
L_{yy} &= \frac{\partial v}{\partial y} = \varepsilon_y \\
L_{zz} &= \frac{\partial w}{\partial z} = \varepsilon_z \\
L_{xy} &= \frac{\partial u}{\partial x} + \frac{\partial v}{\partial x} = \gamma_{xy} \\
L_{yz} &= \frac{\partial w}{\partial y} + \frac{\partial v}{\partial z} = \gamma_{yz} \\
L_{xz} &= \frac{\partial w}{\partial x} + \frac{\partial u}{\partial z} = \gamma_{xz}
\end{aligned}
\qquad (3.18)
$$

Without neglecting the second order terms, the strain components will be expressed as non-linearity relationship, as shown in below:

$$\varepsilon_x \quad = \quad \frac{\partial u}{\partial x} + \frac{1}{2}\left[\left(\frac{\partial u}{\partial x}\right)^2 + \left(\frac{\partial v}{\partial x}\right)^2 + \left(\frac{\partial w}{\partial x}\right)^2\right]$$

$$\varepsilon_y \quad = \quad \frac{\partial v}{\partial y} + \frac{1}{2}\left[\left(\frac{\partial u}{\partial y}\right)^2 + \left(\frac{\partial v}{\partial y}\right)^2 + \left(\frac{\partial w}{\partial y}\right)^2\right]$$

$$\varepsilon_z \quad = \quad \frac{\partial w}{\partial z} + \frac{1}{2}\left[\left(\frac{\partial u}{\partial z}\right)^2 + \left(\frac{\partial v}{\partial z}\right)^2 + \left(\frac{\partial w}{\partial z}\right)^2\right]$$

$$\gamma_{xy} \quad = \quad \frac{\partial u}{\partial y} + \frac{\partial v}{\partial x} + \left(\frac{\partial u}{\partial x}\right) + \left(\frac{\partial u}{\partial y}\right) + \left(\frac{\partial v}{\partial x}\right) + \left(\frac{\partial v}{\partial y}\right) + \left(\frac{\partial w}{\partial x}\right) + \left(\frac{\partial w}{\partial y}\right)$$

$$\gamma_{yz} \quad = \quad \frac{\partial v}{\partial z} + \frac{\partial w}{\partial y} + \left(\frac{\partial u}{\partial y}\right) + \left(\frac{\partial u}{\partial z}\right) + \left(\frac{\partial v}{\partial y}\right) + \left(\frac{\partial v}{\partial z}\right) + \left(\frac{\partial w}{\partial y}\right) + \left(\frac{\partial w}{\partial z}\right)$$

$$\gamma_{xz} \quad = \quad \frac{\partial z}{\partial x} + \frac{\partial u}{\partial z} + \left(\frac{\partial u}{\partial x}\right) + \left(\frac{\partial u}{\partial z}\right) + \left(\frac{\partial v}{\partial x}\right) + \left(\frac{\partial v}{\partial z}\right) + \left(\frac{\partial w}{\partial x}\right) + \left(\frac{\partial w}{\partial z}\right)$$

By substituting Eq. (3.18) into Eq. (3.17) the following is resulted:

$$dL^2 = 2\,d_{x^2}\varepsilon_x + 2\,d_{y^2}\varepsilon_y + 2\,d_{z^2}\varepsilon_z + 2\,d_x d_y \gamma_{xy} + 2\,d_y d_z \gamma_{yz} + 2 d_x d_z \gamma_{xz}$$

3.3   STRAIN COMPATIBILITY EQUATION

With strong attraction force among particles, a body deforms in a stable and predictable manner. This properties of solid makes its deformation exclusive at a specified point. In other words, deformation occurs at any point on a body is dependent of other points. Mathematically, such dependency is expressed in Eqs. (3.2), (3.3), (3.4), (3.6), (3.7) and (3.8).

The change in shear strain γ_{xy} on xy-plane can be expressed by performing partial differentiation to Eq. (3.6):

$$\frac{\partial^2 \gamma_{xy}}{\partial x \partial y} = \frac{\partial^3 u}{\partial x \partial y^2} + \frac{\partial^3 v}{\partial x^2 \partial y} \tag{3.19}$$

The following is produced by performing second order partial differentiation to Eqs. (3.2) and (3.3) with respect to y and x, respectively:

$$\frac{\partial^2 \varepsilon_x}{\partial y^2} = \frac{\partial^3 u}{\partial x \partial y^2}$$

$$\frac{\partial^2 \varepsilon_y}{\partial x^2} = \frac{\partial^3 v}{\partial x^2 \partial y}$$

Substitution of the relationships above in Eq. (3.19) produces the follows:

$$\frac{\partial^2 \gamma_{xy}}{\partial x \partial y} = \frac{\partial^2 \varepsilon_x}{\partial y^2} + \frac{\partial^2 \varepsilon_y}{\partial x^2} \tag{3.20}$$

The equation above is known as compatibility equation, and it must be satisfied to yield consistent displacements. Similarly, compatibility equations for yz and xz planes can be derived as follow:

$$\frac{\partial^2 \gamma_{yz}}{\partial y \partial z} = \frac{\partial^2 \varepsilon_y}{\partial z^2} + \frac{\partial^2 \varepsilon_z}{\partial y^2}$$

$$\frac{\partial^2 \gamma_{xz}}{\partial x \partial z} = \frac{\partial^2 \varepsilon_x}{\partial z^2} + \frac{\partial^2 \varepsilon_z}{\partial x^2}$$

Performing partial differentiation to Eqs. (3.2), (3.3), (3.4), (3.6), (3.7) and (3.8) and the following expressions are resulted:

$$\frac{\partial^2 \varepsilon_x}{\partial y \partial z} = \frac{\partial^3 u}{\partial x \partial y \partial z} \tag{3.21}$$

$$\frac{\partial^2 \varepsilon_y}{\partial x \partial z} = \frac{\partial^3 v}{\partial x \partial y \partial z} \tag{3.22}$$

$$\frac{\partial^2 \varepsilon_z}{\partial x \partial y} = \frac{\partial^3 w}{\partial x \partial y \partial z} \tag{3.23}$$

$$\frac{\partial \gamma_{xy}}{\partial z} = \frac{\partial^2 u}{\partial y \partial z} + \frac{\partial^2 v}{\partial x \partial z} \tag{3.24}$$

$$\frac{\partial \gamma_{yz}}{\partial x} = \frac{\partial^2 v}{\partial x \partial z} + \frac{\partial^2 w}{\partial x \partial y} \tag{3.25}$$

$$\frac{\partial \gamma_{xz}}{\partial y} = \frac{\partial^2 u}{\partial y \partial z} + \frac{\partial^2 w}{\partial x \partial y} \tag{3.26}$$

By performing partial differentiation to Eq. (3.24) with respect to x yields the following:

$$\frac{\partial^2 \gamma_{xy}}{\partial x \partial z} = \frac{\partial^3 u}{\partial x \partial y \partial z} + \frac{\partial^3 v}{\partial x^2 \partial z} \tag{3.27}$$

Partial differentiation of Eq. (3.26) with respect to x will produce the following:

$$\frac{\partial^2 \gamma_{xz}}{\partial x \partial y} = \frac{\partial^3 u}{\partial x \partial y \partial z} + \frac{\partial^3 w}{\partial x^2 \partial y} \tag{3.28}$$

By adding $\dfrac{\partial^2 \gamma_{xy}}{\partial x \partial z}$ and $\dfrac{\partial^2 \gamma_{xz}}{\partial x \partial y}$ in Eqs. (3.27) and (3.28) and we get this:

$$\frac{\partial^2 \gamma_{xy}}{\partial x \partial z} + \frac{\partial^2 \gamma_{xz}}{\partial x \partial y} = \left(\frac{\partial^3 u}{\partial x \partial y \partial z} + \frac{\partial^3 v}{\partial x^2 \partial z} \right) + \left(\frac{\partial^3 u}{\partial x \partial y \partial z} + \frac{\partial^3 w}{\partial x^2 \partial y} \right)$$

The following is obtained through simplification of equation above:

$$\frac{\partial}{\partial x}\left(\frac{\partial \gamma_{xy}}{\partial z} + \frac{\partial \gamma_{xz}}{\partial y} \right) = \frac{\partial^2}{\partial x \partial z}\left(\frac{\partial u}{\partial y} + \frac{\partial v}{\partial x} \right) + \frac{\partial^2}{\partial x \partial y}\left(\frac{\partial u}{\partial z} + \frac{\partial w}{\partial x} \right)$$

Rearrangement of the equation above results the follows:

$$\frac{\partial}{\partial x}\left(\frac{\partial \gamma_{xy}}{\partial z} + \frac{\partial \gamma_{xz}}{\partial y} \right) = \frac{\partial^2}{\partial y \partial z}\left(\frac{\partial u}{\partial x} \right) + \frac{\partial^2}{\partial x^2}\left(\frac{\partial v}{\partial z} + \frac{\partial w}{\partial y} \right) + \frac{\partial^2}{\partial y \partial z}\left(\frac{\partial u}{\partial x} \right)$$

By substituting Eqs. (3.2) and (3.8) into equation above yields the following equation:

$$\frac{\partial}{\partial x}\left(\frac{\partial \gamma_{xy}}{\partial z} + \frac{\partial \gamma_{xz}}{\partial y} \right) = \frac{\partial^2}{\partial y \partial z}\varepsilon_x + \frac{\partial^2}{\partial x^2}\gamma_{yz} + \frac{\partial^2}{\partial y \partial z}\varepsilon_x$$

The following equation can be written by rearranging the equation above:

$$2\frac{\partial^2}{\partial y \partial z}\varepsilon_x = \frac{\partial}{\partial x}\left(-\frac{\partial \gamma_{yz}}{\partial x} + \frac{\partial \gamma_{xz}}{\partial y} + \frac{\partial \gamma_{xy}}{\partial z} \right) \qquad (3.29)$$

By performing partial differentiation to Eq. (3.24) with respect to y yields the following:

$$\frac{\partial^2 \gamma_{xy}}{\partial y \partial z} = \frac{\partial^3 u}{\partial y^2 \partial z} + \frac{\partial^3 v}{\partial x \partial y \partial z} \qquad (3.30)$$

Partial differentiation of Eq. (3.25) with respect to y results the following:

$$\frac{\partial^2 \gamma_{yz}}{\partial x \partial y} = \frac{\partial^3 v}{\partial x \partial y \partial z} + \frac{\partial^3 w}{\partial x \partial y^2} \qquad (3.31)$$

By adding $\dfrac{\partial^2 \gamma_{xy}}{\partial y \partial z}$ and $\dfrac{\partial^2 \gamma_{yz}}{\partial x \partial y}$ in Eqs. (3.30) and (3.31) gives us the following:

$$\frac{\partial^2 \gamma_{xy}}{\partial y \partial z} + \frac{\partial^2 \gamma_{yz}}{\partial x \partial y} = \left(\frac{\partial^3 u}{\partial y^2 \partial z} + \frac{\partial^3 v}{\partial x \partial y \partial z} \right) + \left(\frac{\partial^3 v}{\partial x \partial y \partial z} + \frac{\partial^3 w}{\partial x \partial y^2} \right)$$

By simplifying the equation, we can get:

$$\frac{\partial}{\partial y} \left(\frac{\partial \gamma_{xy}}{\partial z} + \frac{\partial \gamma_{yz}}{\partial x} \right) = \frac{\partial^2}{\partial y \partial z} \left(\frac{\partial u}{\partial y} + \frac{\partial v}{\partial x} \right) + \frac{\partial^2}{\partial x \partial y} \left(\frac{\partial v}{\partial z} + \frac{\partial w}{\partial y} \right)$$

By rearranging the equation above the following is obtained:

$$\frac{\partial}{\partial y} \left(\frac{\partial \gamma_{xy}}{\partial z} + \frac{\partial \gamma_{yz}}{\partial x} \right) = \frac{\partial^2}{\partial x \partial z} \left(\frac{\partial v}{\partial y} \right) + \frac{\partial^2}{\partial y^2} \left(\frac{\partial u}{\partial z} + \frac{\partial w}{\partial x} \right) + \frac{\partial^2}{\partial x \partial z} \left(\frac{\partial v}{\partial y} \right)$$

Substitution of Eqs. (3.3) and (3.7) into equation above produces the following:

$$\frac{\partial}{\partial y} \left(\frac{\partial \gamma_{xy}}{\partial z} + \frac{\partial \gamma_{yz}}{\partial x} \right) = \frac{\partial^2}{\partial x \partial z} \varepsilon_y + \frac{\partial^2}{\partial y^2} \gamma_{xz} + \frac{\partial^2}{\partial x \partial z} \varepsilon_y$$

With rearrangement the following equation is resulted:

$$2 \frac{\partial^2}{\partial x \partial z} \varepsilon_y = \frac{\partial}{\partial y} \left(\frac{\partial \gamma_{yz}}{\partial x} - \frac{\partial \gamma_{xz}}{\partial y} + \frac{\partial \gamma_{xy}}{\partial z} \right) \qquad (3.32)$$

Partial differentiation of Eq. (3.24) with respect to z leads to the following:

$$\frac{\partial^2 \gamma_{yz}}{\partial x \partial z} = \frac{\partial^3 v}{\partial x \partial z^2} + \frac{\partial^3 w}{\partial x \partial y \partial z} \qquad (3.33)$$

By performing partial differentiation to Eq. (3.26) with respect to z yields the following:

$$\frac{\partial^2 \gamma_{xz}}{\partial y \partial z} = \frac{\partial^3 u}{\partial y \partial z^2} + \frac{\partial^3 w}{\partial x \partial y \partial z} \qquad (3.34)$$

Addition of $\frac{\partial^2 \gamma_{yz}}{\partial y \partial z}$ and $\frac{\partial^2 \gamma_{xz}}{\partial x \partial y}$ in Eqs. (3.33) and (3.24) produces the following equation:

$$\frac{\partial^2 \gamma_{yz}}{\partial x \partial z} + \frac{\partial^2 \gamma_{xz}}{\partial y \partial z} = \left(\frac{\partial^3 v}{\partial x \partial z^2} + \frac{\partial^3 w}{\partial x \partial y \partial z} \right) + \left(\frac{\partial^3 u}{\partial y \partial z^2} + \frac{\partial^3 w}{\partial x \partial y \partial z} \right)$$

Through simplification of the equation above we can get:

$$\frac{\partial}{\partial z}\left(\frac{\partial \gamma_{yz}}{\partial x} + \frac{\partial \gamma_{xz}}{\partial y}\right) = \frac{\partial^2}{\partial x \partial z}\left(\frac{\partial v}{\partial z} + \frac{\partial w}{\partial y}\right) + \frac{\partial^2}{\partial y \partial z}\left(\frac{\partial u}{\partial z} + \frac{\partial w}{\partial x}\right)$$

Rearrange the equation above yields the follows:

$$\frac{\partial}{\partial z}\left(\frac{\partial \gamma_{yz}}{\partial x} + \frac{\partial \gamma_{xz}}{\partial y}\right) = \frac{\partial^2}{\partial x \partial y}\left(\frac{\partial w}{\partial y}\right) + \frac{\partial^2}{\partial z^2}\left(\frac{\partial v}{\partial x} + \frac{\partial u}{\partial y}\right) + \frac{\partial^2}{\partial x \partial z}\left(\frac{\partial w}{\partial y}\right)$$

The equation below is obtained after substituting Eqs. (3.4) and (3.6) into equation above:

$$\frac{\partial}{\partial z}\left(\frac{\partial \gamma_{yz}}{\partial x} + \frac{\partial \gamma_{xz}}{\partial y}\right) = \frac{\partial^2}{\partial x \partial z}\varepsilon_z + \frac{\partial^2}{\partial z^2}\gamma_{xy} + \frac{\partial^2}{\partial x \partial z}\varepsilon_z$$

Rearrange the equation above results in the following expression:

$$2\frac{\partial^2}{\partial x \partial z}\varepsilon_z = \frac{\partial}{\partial z}\left(\frac{\partial \gamma_{yz}}{\partial x} + \frac{\partial \gamma_{xz}}{\partial y} - \frac{\partial \gamma_{xy}}{\partial z}\right) \qquad (3.35)$$

3.4   STRAIN TRANSFORMATION

Say the direction cosines for two different coordinate systems are as shown in Table 3.1:

Corresponding to stress components in the form of Eq. (2.31), the transformation of strain components in Eq. (3.14) can be derived as follows:

$$\begin{bmatrix} \varepsilon_{x'} & \frac{1}{2}\gamma_{y'x'} & \frac{1}{2}\gamma_{z'x'} \\ \frac{1}{2}\gamma_{x'y'} & \varepsilon_{y'} & \frac{1}{2}\gamma_{z'y'} \\ \frac{1}{2}\gamma_{x'z'} & \frac{1}{2}\gamma_{y'z'} & \varepsilon_{z''} \end{bmatrix} = \begin{bmatrix} l_1 m_1 n_1 \\ l_2 m_2 n_2 \\ l_3 m_3 n_3 \end{bmatrix} \begin{bmatrix} \varepsilon_x & \frac{1}{2}\gamma_{yx} & \frac{1}{2}\gamma_{zx} \\ \frac{1}{2}\gamma_{xy} & \varepsilon_y & \frac{1}{2}\gamma_{zy} \\ \frac{1}{2}\gamma_{xz} & \frac{1}{2}\gamma_{yz} & \varepsilon_z \end{bmatrix} \begin{bmatrix} l_1 l_2 l_3 \\ m_1 m_2 m_3 \\ n_1 n_2 n_3 \end{bmatrix} \qquad (3.36)$$

TABLE 3.1

Direction cosines between new and old coordinate systems

	x	y	z
x'	l_1	m_1	n_1
y'	l_2	m_2	n_2
z'	l_3	m_3	n_3

3.5 PRINCIPAL STRAIN AND MAXIMUM SHEAR STRAIN

Corresponding to stress components in the form of Eq. (2.31), the corresponding strain components in Eq. (3.14) are applied for the strain invariants. By replacing the stress components with corresponding strain components as per Eq. (3.14) into Eq. (2.35) leads to the follows:

$$I_1 = \epsilon_x + \epsilon_y + \epsilon_z \tag{3.37}$$

By replacing the stress components with corresponding strain components as per Eq. (3.14) into Eq. (2.36) results in the following:

$$I_2 = \epsilon_x \epsilon_y + \epsilon_y \epsilon_z + \epsilon_x \epsilon_z - \left(\frac{1}{2}\gamma_{xy}\right)^2 - \left(\frac{1}{2}\gamma_{yz}\right)^2 - \left(\frac{1}{2}\gamma_{xz}\right)^2$$

The following yielded after expansion of equation above:

$$I_2 = \epsilon_x \epsilon_y + \epsilon_y \epsilon_z + \epsilon_x \epsilon_z - \frac{1}{4}\gamma_{xy}^2 - \frac{1}{4}\gamma_{yz}^2 - \frac{1}{4}\gamma_{xz}^2$$

Through simplification we can get the following equation:

$$I_2 = \epsilon_x \epsilon_y + \epsilon_y \epsilon_z + \epsilon_x \epsilon_z - \frac{1}{4}\left(\gamma_{xy}^2 + \gamma_{yz}^2 + \gamma_{xz}^2\right)$$

By replacing the stress components with corresponding strain components as per Eq. (3.14) into Eq. (2.37) yields the follows:

$$I_3 = \epsilon_x \epsilon_y \epsilon_z - \epsilon_x \left(\frac{1}{2}\gamma_{yz}\right)^2 - \epsilon_y \left(\frac{1}{2}\gamma_{xz}\right)^2 - \epsilon_z \left(\frac{1}{2}\gamma_{xy}\right)^2 + 2\left(\frac{1}{2}\gamma_{xy}\right)\left(\frac{1}{2}\gamma_{yz}\right)\left(\frac{1}{2}\gamma_{xz}\right)$$

Expand the equation above results in the following expression:

$$I_3 = \epsilon_x \epsilon_y \epsilon_z - \epsilon_x \frac{1}{4}\gamma_{yz}^2 - \epsilon_y \frac{1}{4}\gamma_{xz}^2 - \epsilon_z \frac{1}{4}\gamma_{xy}^2 + \frac{1}{4}\left(\gamma_{xy}\gamma_{yz}\gamma_{xz}\right)$$

Simplify the equation above and we obtain:

$$I_3 = \epsilon_x \epsilon_y \epsilon_z - \frac{1}{4}\left(\epsilon_x \gamma_{yz}^2 + \epsilon_y \gamma_{xz}^2 + \epsilon_z \gamma_{xy}^2\right) + \frac{1}{4}\left(\gamma_{xy}\gamma_{yz}\gamma_{xz}\right)$$

The maximum shear strain can be expressed by replacing the stress components with corresponding strain components as per Eq. (3.14) into relationships in Eq. (2.44). Say, for $\gamma_{1,2,\,max}$:

$$\frac{1}{2}\gamma_{1,2,\,max} = \pm\frac{(\epsilon_1 - \epsilon_2)}{2}$$

Simplify the equation above yields the follow:

$$\gamma_{1,2,\,max} = \pm\epsilon_1 - \epsilon_2$$

Similarly, shear strain for other planes can be written in the following form:

$$\gamma_{2,3,\,max} = \pm\epsilon_2 - \epsilon_3$$

$$\gamma_{1,3,\,max} = \pm\epsilon_1 - \epsilon_3$$

3.6   DEVIATORIC STRAIN

Like stresses, the deviatoric strain can be written with the corresponding strain components in Eq. (3.14). Replacing the stress components with those corresponding strain components in Eq. (2.45) yields the following:

$$\epsilon = \frac{\epsilon_x + \epsilon_y + \epsilon_z}{3}$$

Substitution of Eq. (3.37) into equation above provides us the definition below:

$$\epsilon = \frac{I_1}{3}$$

The deviatoric strain invariants in the other hand, remains the same as they are for the stress components:

$$J_2 = \frac{I_1^2}{3} - I_2$$

$$J_3 = I_3 - I_2\sigma_m + 2\sigma_m^3$$

3.7 OCTAHEDRAL STRAIN

The octahedral strains are no different from the octahedral stresses in term of the structure of expression. The following expression is obtained by replacing the stress components with the corresponding strain components as per Eq. (3.14) in Eq. (2.52):

$$\epsilon_{oct} = \frac{\epsilon_1 + \epsilon_2 + \epsilon_3}{3}$$

Similarly, octahedral shear strain can be derived by replacing the stress components with the corresponding strain components as per Eq. (3.14) in Eq. (2.54):

$$\frac{1}{2}\gamma_{oct} = \pm\frac{1}{3}\sqrt{(\epsilon_1 - \epsilon_2)^2 + (\epsilon_2 - \epsilon_3)^2 + (\epsilon_3 - \epsilon_1)^2}$$

The following equation is resulted after simplification of equation above:

$$\gamma_{oct} = \pm\frac{2}{3}\sqrt{(\epsilon_1 - \epsilon_2)^2 + (\epsilon_2 - \epsilon_3)^2 + (\epsilon_3 - \epsilon_1)^2}$$

3.8 PLANE STRAIN

Consider the length of a body (say, along z-axis) is much greater than its width and height (along x and y axes). When force exerted in the direction normal to z-axis, strains are developed in any direction but along z-axis. Since there is no longitudinal strain, such scenario can be simplified into plane strain scenario, as shown in Fig. 3.5.

Under plane strain scenario, the component of strain as per Eq. (3.13) can be simplified as follows:

$$[\epsilon] = \begin{bmatrix} \epsilon_x & \gamma_{yx} & 0 \\ \gamma_{xy} & \epsilon_y & 0 \\ 0 & 0 & 0 \end{bmatrix}$$

In expression above, it is noteworthy that γ_{yx} and γ_{xy} are only due to corresponding shear stress components, τ_{yx} and τ_{xy}.

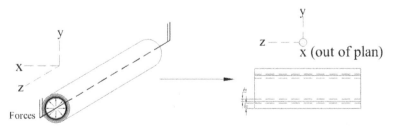

FIGURE 3.5 Plane strain.

4 Stress–Strain Relationships

4.1 TYPES OF RELATIONSHIPS

Solids are full of gaps between particles by nature. When force is exerted, the particles at the point of exertion are forced to displace. In the presence of a force of attraction, the other particles are pulled towards them, filling the gaps nearby.

When force is removed, the attraction force will pull all the displaced particles back to their original position as that position forms the most stable arrangement in solids. When all particles are back to their original position, there will be no deformation left with respect to the original solid. Under this condition, the solid exhibits an elastic behaviour, as demonstrated in Fig. 4.1.

When the force exerted is too strong, it may overcome the attraction force. In this case, the unlinked particles can only be pulled or pushed towards their original position indirectly upon removal of the exerted force. In most situations, they will fail to return to their original position and other particles will often slide over each other due to a unbalanced force of attraction. As a result, deformation with respect to the original solid can be observed even when there is no external force. Under this condition, the solid exhibits plastic behaviour, as shown in Fig. 4.2.

The stress–strain relationship is usually determined through experiment. The result is plotted on a graph of stress against strain. This graph clearly shows the behaviour of the material under different amounts of stresses, and thus enables us to determine its yield strength for engineering purposes. For example, the stress–strain curve of steel is shown in Fig. 4.3.

From Fig. 4.3, point A is identified as the elastic limit and the corresponding stress is known as yield stress, σ_y. Prior to point A, the body behaviour obeys Hooke's Law, in which the strain is solely dependent on stress. At point A, increase in stress will cause the body to yield (loss of some attraction forces) and exhibit plastic behaviour. Unlike elastic behaviour, plastic behaviour is unpredictable as the particles do not always move in the same way after they lose their attraction force. Nevertheless, the body can still resist higher stress because most of the particles are still intact. Point B is an optimum point where the ultimate stress resistance, σ_{ult}, is achieved in the midst of continuous loss of attraction force due to increasing strain and stress. Beyond that point, particles in the body literally break apart and fracture occurs at point C.

Modulus of elasticity is defined as the amount of stress required to produce one unit of strain and measures the rigidity of a body when it is subjected to uniaxial tension or compression while exhibiting elastic behaviour. It can be easily obtained from the graph of stress against strain by calculating its gradient under elastic zone (OA).

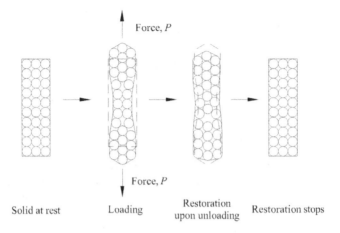

FIGURE 4.1   Elastic behaviour of a body at the microscopic level.

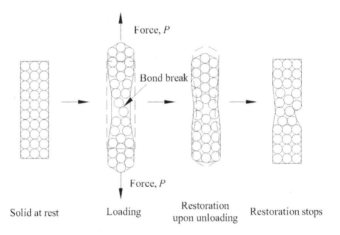

FIGURE 4.2   Plastic behaviour of a body at microscopic level.

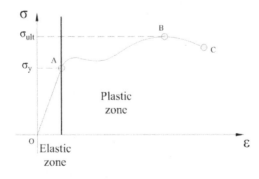

FIGURE 4.3   Stress–strain curve for steel.

$$Modulus \ of \ Elasticity, \ E = \frac{\sigma}{\epsilon} \qquad (4.1)$$

By substituting Eqs. (2.1) and (3.1) into Eq. (4.1), the expression for modulus of elasticity is transformed to:

$$E = \frac{F}{A} \times \frac{L}{\Delta L} \qquad (4.2)$$

Modulus of elasticity provides insights into the strength of attraction force between particles. This attraction force acts in all directions. Even when the force exerted at one point in one direction, the attraction force will pull and push all particles around it. Therefore, a body is expected to develop stress and deform in all directions even though the force is exerted in one direction only. The ratio of transverse strain to longitudinal strain is known as Poisson's ratio, as defined in Fig. 4.4.

$$Poisson's \ ratio, \ \nu = -\frac{\epsilon_{trans}}{\epsilon_{long}} \qquad (4.3)$$

In the above expression, ϵ_{trans} is transverse strain, which defines the deformation normal to the direction of force. On the other hand, ϵ_{long} is longitudinal strain, which defines the deformation along the direction of the applied force.

Shear modulus measures a body's shear rigidity, which has a similar context as modulus of elasticity. Likewise, a greater value in shear modulus denotes that the body does not deform in the direction of shear stress easily.

$$Shear \ modulus, \ \ G = \frac{\tau}{\gamma} \qquad (4.4)$$

where

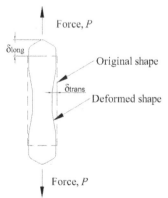

FIGURE 4.4   Longitudinal and transverse deformation of a body.

τ is shear stress;
γ is shear strain.

4.2   GENERALISED STRESS–STRAIN RELATIONSHIP

From Eq. (4.1), stress can be expressed in terms of the following relationship:

$$\sigma = E\epsilon$$

Since σ and ϵ are stress and strain components at a point in matrix form, E would be expressed in matrix form as well to generalise the stress–strain relationship. In this case, D matrix is introduced.

$$\{\sigma\} = [D]\{\epsilon\}$$

Based on the relationships as shown in Eqs. (2.8), (2.9) and (2.10), we know that the following shear stress components are complementary:

$$\tau_{yz} = \tau_{zy}$$

$$\tau_{xy} = \tau_{yx}$$

$$\tau_{zx} = \tau_{xz}$$

Therefore, the corresponding strain components have the same relationship:

$$\gamma_{yz} = \gamma_{zy}$$

$$\gamma_{xy} = \gamma_{yx}$$

$$\gamma_{zx} = \gamma_{xz}$$

Therefore, the generalised stress–strain relationship for anisotropic material for stress and strain components as stated in Eq. (2.11) and Eq. (3.36), respectively, can be expressed in the form of $\{\sigma\} = [D]\{\epsilon\}$:

$$\begin{bmatrix} \sigma_x \\ \sigma_y \\ \sigma_z \\ \tau_{xy} \\ \tau_{yx} \\ \tau_{xz} \end{bmatrix} = \begin{bmatrix} D_{11} & D_{12} & D_{13} & D_{14} & D_{15} & D_{16} \\ D_{21} & D_{22} & D_{23} & D_{24} & D_{25} & D_{26} \\ D_{31} & D_{32} & D_{33} & D_{34} & D_{35} & D_{36} \\ D_{41} & D_{42} & D_{43} & D_{44} & D_{45} & D_{46} \\ D_{51} & D_{52} & D_{53} & D_{54} & D_{55} & D_{56} \\ D_{61} & D_{62} & D_{63} & D_{64} & D_{65} & D_{66} \end{bmatrix} \begin{bmatrix} \epsilon_x \\ \epsilon_y \\ \epsilon_z \\ \gamma_{xy} \\ \gamma_{yx} \\ \gamma_{xz} \end{bmatrix} \qquad (4.5)$$

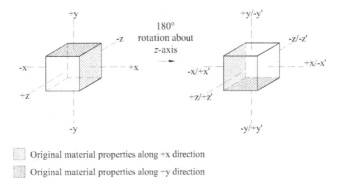

FIGURE 4.5   Solid before and after 180° rotation about z-axis.

D matrix can be derived from Eq. (4.5) for any type of material by taking an anisotropic material as the reference material. Sections 4.3 to 4.7 illustrate progressive derivation of D matrix for various materials sorted in a descending order of the level of anisotropy.

4.3   MATERIAL WITH SYMMETRICAL PROPERTIES ABOUT Z AXIS

Say there is a material with properties symmetrical about the z axis, which means the material properties along the positive x-direction are the same as those in the negative x-direction, and so on, for y-axis. By rotating the solid by 180° about the z-axis, the face of the solid that carries the properties of material in the positive x-direction now coincides with the face that carries properties in the negative x-direction as per the global coordinate system, which is the coordinate system before rotation, as shown in Fig. 4.5.

The direction cosines between mutually orthogonal axes before and after rotation are shown in Table 4.1.

The following stress and strain expressions are obtained by substituting the values in Table 4.1 to Eqs. (2.31) and (3.36):

TABLE 4.1
Transformation of axis for material with symmetrical properties about z axis

	x	Y	z
x'	$\cos 180° = -1$	$\cos 90° = 0$	$\cos 90° = 0$
y'	$\cos 90° = 0$	$\cos 180° = -1$	$\cos 90° = 0$
z'	$\cos 90° = 0$	$\cos 90° = 0$	$\cos 0° = 1$

$$
\begin{bmatrix} \sigma_{x'}\tau_{x'y'}\tau_{x'z'} \\ \tau_{y'x'}\sigma_{y'}\tau_{y'z'} \\ \tau_{z'x'}\tau_{z'y'}\sigma_{z'} \end{bmatrix} = \begin{bmatrix} -1 & 0 & 0 \\ 0 & -1 & 0 \\ 0 & 0 & 1 \end{bmatrix} \begin{bmatrix} \sigma_x\tau_{xy}\tau_{xz} \\ \tau_{yx}\sigma_y\tau_{yz} \\ \tau_{zx}\tau_{zy}\sigma_z \end{bmatrix} \begin{bmatrix} -1 & 0 & 0 \\ 0 & -1 & 0 \\ 0 & 0 & 1 \end{bmatrix} = \begin{bmatrix} \sigma_x & \tau_{xy} & -\tau_{xz} \\ \tau_{yx} & \sigma_y & -\tau_{yz} \\ -\tau_{zx} & -\tau_{zy} & \sigma_z \end{bmatrix}
$$

$$(4.6)$$

$$
\begin{bmatrix} \epsilon_{x'} & \frac{1}{2}\gamma_{y'x'} & \frac{1}{2}\gamma_{z'x'} \\ \frac{1}{2}\gamma_{x'y'} & \epsilon_{y'} & \frac{1}{2}\gamma_{z'y'} \\ \frac{1}{2}\gamma_{x'z'} & \frac{1}{2}\gamma_{y'z'} & \epsilon_{z'} \end{bmatrix} = \begin{bmatrix} -1 & 0 & 0 \\ 0 & -1 & 0 \\ 0 & 0 & 1 \end{bmatrix} \begin{bmatrix} \epsilon_x & \frac{1}{2}\gamma_{yx} & \frac{1}{2}\gamma_{zx} \\ \frac{1}{2}\gamma_{xy} & \epsilon_y & \frac{1}{2}\gamma_{zy} \\ \frac{1}{2}\gamma_{xz} & \frac{1}{2}\gamma_{yz} & \epsilon_z \end{bmatrix} \begin{bmatrix} -1 & 0 & 0 \\ 0 & -1 & 0 \\ 0 & 0 & 1 \end{bmatrix}
$$

$$
= \begin{bmatrix} \epsilon_x & \frac{1}{2}\gamma_{xy} & -\frac{1}{2}\gamma_{xz} \\ \frac{1}{2}\gamma_{yx} & \epsilon_y & -\frac{1}{2}\gamma_{yz} \\ -\frac{1}{2}\gamma_{zx} & -\frac{1}{2}\gamma_{zy} & \epsilon_z \end{bmatrix}
$$

$$(4.7)$$

By comparing the stress and strain components before and after transformation in Eqs. (4.6) and (4.7), respectively, we get the following relationship:

$$
\begin{aligned}
\sigma_{x'} &= \sigma_x \tau_{x'y'} = \tau_{xy} \\
\sigma_{y'} &= \sigma_y \tau_{x'z'} = -\tau_{xz} \\
\sigma_{z'} &= \sigma_z \tau_{y'z'} = -\tau_{yz} \\
\epsilon_{x'} &= \epsilon_x \gamma_{x'y'} = \gamma_{xy} \\
\epsilon_{y'} &= \epsilon_y \gamma_{x'z'} = -\gamma_{xz} \\
\epsilon_{z'} &= \epsilon_z \gamma_{y'z'} = -\gamma_{yz}
\end{aligned}
$$

$$(4.8)$$

By referring to Eq. (4.5), σ_x before rotation can be expressed as follows:

$$\sigma_x = D_{11}\epsilon_x + D_{12}\epsilon_y + D_{13}\epsilon_z + D_{14}\gamma_{xy} + D_{15}\gamma_{yz} + D_{16}\gamma_{xz} \qquad (4.9)$$

The expression for $\sigma_{x'}$, which denotes the normal stress in the positive x-direction after rotation, can be written in the following form based on the general form as shown in Eq. (4.9):

$$\sigma_{x'} = D_{11}\epsilon_{x'} + D_{12}\epsilon_{y'} + D_{13}\epsilon_{z'} + D_{14}\gamma_{x'y'} + D_{15}\gamma_{y'z'} + D_{16}\gamma_{x'z'} \qquad (4.10)$$

By equating the normal stress in the positive x-direction before and after rotation, we apply the condition where the stresses to be developed in both positive and negative directions along the original global x-axis are the same. This simulates the case where the material properties are the same for the faces normal to x-axis, as shown in Fig. 4.6.

However, Eq. (4.10) needs to be expressed as σ_x in order to do so. This can be achieved by applying the relationships as shown in Eq. (4.8):

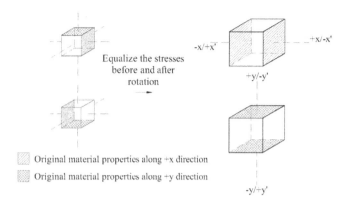

FIGURE 4.6 Transformation from anisotropic material to material with symmetrical properties about z-axis.

$$\sigma_x = D_{11}\varepsilon_x + D_{12}\varepsilon_y + D_{13}\varepsilon_z + D_{14}\gamma_{xy} - D_{15}\gamma_{yz} - D_{16}\gamma_{xz} \qquad (4.11)$$

By equating the terms on the right-hand side in Eqs. (4.9) and (4.11) the following is obtained:

$$D_{11}\varepsilon_x + D_{12}\varepsilon_y + D_{13}\varepsilon_z + D_{14}\gamma_{xy} + D_{15}\gamma_{yz} + D_{16}\gamma_{xz}$$
$$= D_{11}\varepsilon_x + D_{12}\varepsilon_y + D_{13}\varepsilon_z + D_{14}\gamma_{xy} - D_{15}\gamma_{yz} - D_{16}\gamma_{xz}$$

By comparing the coefficients of the common terms in the equation above, we know that

$$D_{15} = -D_{15}$$

$$D_{16} = -D_{16}$$

The equations above are valid if and only if both D_{15} and D_{16} are zero. The other D components remain the same as they are for anisotropic material.

By referring to Eq. (4.5), σ_y before rotation can be expressed as follows:

$$\sigma_y = D_{21}\varepsilon_x + D_{22}\varepsilon_y + D_{23}\varepsilon_z + D_{24}\gamma_{xy} + D_{25}\gamma_{yz} + D_{26}\gamma_{xz} \qquad (4.12)$$

The expression for $\sigma_{y'}$ can be written as below, based on the general form as shown in Eq. (4.12):

$$\sigma_{y'} = D_{21}\varepsilon_{x'} + D_{22}\varepsilon_{y'} + D_{23}\varepsilon_{z'} + D_{24}\gamma_{x'y'} + D_{25}\gamma_{y'z'} + D_{26}\gamma_{x'z'} \qquad (4.13)$$

The following is produced after applying relationship in Eq. (4.8) to the equation above:

$$\sigma_y = D_{21}\varepsilon_x + D_{22}\varepsilon_y + D_{23}\varepsilon_z + D_{24}\gamma_{xy} - D_{25}\gamma_{yz} - D_{26}\gamma_{xz} \qquad (4.14)$$

Similarly, by equating the terms on the right-hand side in Eqs. (4.12) and (4.14) and then comparing the coefficients of the common terms, we know that:

$$D_{25} = -D_{25}$$

$$D_{26} = -D_{26}$$

The equations above are valid if and only if both D_{25} and D_{26} are zero. The other D components remain the same as they are for anisotropic material.

From the derivation above, we can summarise that the coefficients of σ_x and σ_y (e.g. D_{15} and D_{16}) are zero when stresses before are the same as after rotation (e.g. $\sigma_x = \sigma'_x$), while the strains associated with those coefficients are equal to their counterparts in opposite sign (e.g. $\gamma_{y'z'} = -\gamma_{yz}$ and $\gamma_{x'z'} = -\gamma_{xz}$). Therefore, D_{35}, D_{36}, D_{45} and D_{46} are zero because of the following expressions:

$$\sigma_z = D_{31}\varepsilon_x + D_{32}\varepsilon_y + D_{33}\varepsilon_z + D_{34}\gamma_{xy} + D_{35}\gamma_{yz} + D_{36}\gamma_{xz}$$

$$\sigma_z = \sigma_{z'},\ \gamma_{y'z'} = -\gamma_{yz}\ and\ \gamma_{x'z'} = -\gamma_{xz}$$

$$\tau_{xy} = D_{41}\varepsilon_x + D_{42}\varepsilon_y + D_{43}\varepsilon_z + D_{44}\gamma_{xy} + D_{45}\gamma_{yz} + D_{46}\gamma_{xz}$$

$$\tau_{xy} = \tau_{x'y'},\ \gamma_{y'z'} = -\gamma_{yz}\ and\ \gamma_{x'z'} = -\gamma_{xz}$$

By referring to Eq. (4.5), τ_{yz} before rotation can be expressed as:

$$\tau_{yz} = D_{51}\varepsilon_x + D_{52}\varepsilon_y + D_{53}\varepsilon_z + D_{54}\gamma_{xy} + D_{55}\gamma_{yz} + D_{56}\gamma_{xz} \qquad (4.15)$$

The expression for $\tau_{y'z'}$ can be written based on the general form as shown in Eq. (4.15):

$$\tau_{y'z'} = D_{51}\varepsilon_{x'} + D_{52}\varepsilon_{y'} + D_{53}\varepsilon_{z'} + D_{54}\gamma_{x'y'} + D_{55}\gamma_{y'z'} + D_{56}\gamma_{x'z'} \qquad (4.16)$$

Applying the relationship in Eq. (4.8) to the equation above results in the following equations:

$$- \tau_{yz} = D_{51}\varepsilon_x + D_{52}\varepsilon_y + D_{53}\varepsilon_z + D_{54}\gamma_{xy} - D_{55}\gamma_{yz} - D_{56}\gamma_{xz}$$

$$\tau_{yz} = -D_{51}\varepsilon_x - D_{52}\varepsilon_y - D_{53}\varepsilon_z - D_{54}\gamma_{xy} + D_{55}\gamma_{yz} + D_{56}\gamma_{xz} \tag{4.17}$$

By equating the terms on the right-hand side in Eqs. (4.15) and (4.17) we obtain the following:

$$D_{51}\varepsilon_x + D_{52}\varepsilon_y + D_{53}\varepsilon_z + D_{54}\gamma_{xy} + D_{55}\gamma_{yz} + D_{56}\gamma_{xz}$$
$$= - D_{51}\varepsilon_x - D_{52}\varepsilon_y - D_{53}\varepsilon_z - D_{54}\gamma_{xy} + D_{55}\gamma_{yz} + D_{56}\gamma_{xz}$$

The following is obtained by comparing the coefficients of the common terms in equation above:

$$D_{51} = -D_{51}$$

$$D_{52} = -D_{52}$$

$$D_{53} = -D_{53}$$

$$D_{54} = -D_{54}$$

The equations above are valid if and only if D_{51}, D_{52}, D_{53} and D_{54} are zero. The other D components remain the same as they are for anisotropic material.

By referring to Eq. (4.5), τ_{xz} before rotation can be expressed as below:

$$\tau_{xz} = D_{61}\varepsilon_x + D_{62}\varepsilon_y + D_{63}\varepsilon_z + D_{64}\gamma_{xy} + D_{65}\gamma_{yz} + D_{66}\gamma_{xz} \tag{4.18}$$

Based on the general form as shown in Eq. (4.18), the expression for $\tau_{x'z'}$ can be written as below:

$$\tau_{x'z'} = D_{61}\varepsilon_{x'} + D_{62}\varepsilon_{y'} + D_{63}\varepsilon_{z'} + D_{64}\gamma_{x'y'} + D_{65}\gamma_{y'z'} + D_{66}\gamma_{x'z'} \tag{4.19}$$

Application of relationship in Eq. (4.8) to the equation above produces the following:

$$- \tau_{xz} = D_{61}\varepsilon_x + D_{62}\varepsilon_y + D_{63}\varepsilon_z + D_{64}\gamma_{xy} - D_{65}\gamma_{yz} - D_{66}\gamma_{xz}$$

$$\tau_{xz} = -D_{61}\varepsilon_x - D_{62}\varepsilon_y - D_{63}\varepsilon_z - D_{64}\gamma_{xy} + D_{65}\gamma_{yz} + D_{66}\gamma_{xz} \tag{4.20}$$

Equating the terms on the right-hand side in Eqs. (4.18) and (4.20), and comparing the coefficients of the common terms leads to the following:

$$D_{61} = -D_{61}$$

$$D_{62} = -D_{62}$$

$$D_{63} = -D_{63}$$

$$D_{64} = -D_{64}$$

The equations above are valid if and only if D_{61}, D_{62}, D_{63} and D_{64} are zero. The other D components remain the same as they are for anisotropic material.

With all the derivation above, the expression in Eq. (4.5) is simplified as follow for material with symmetrical properties along the z axis:

$$\begin{bmatrix} \sigma_x \\ \sigma_y \\ \sigma_z \\ \tau_{xy} \\ \tau_{yz} \\ \tau_{zx} \end{bmatrix} = \begin{bmatrix} D_{11} & D_{12} & D_{13} & D_{14} & 0 & 0 \\ D_{21} & D_{22} & D_{23} & D_{24} & 0 & 0 \\ D_{31} & D_{32} & D_{33} & D_{34} & 0 & 0 \\ D_{41} & D_{42} & D_{43} & D_{44} & 0 & 0 \\ 0 & 0 & 0 & 0 & D_{55} & D_{56} \\ 0 & 0 & 0 & 0 & D_{65} & D_{66} \end{bmatrix} \begin{bmatrix} \varepsilon_x \\ \varepsilon_y \\ \varepsilon_z \\ \gamma_{xy} \\ \gamma_{yz} \\ \gamma_{zx} \end{bmatrix} \quad (4.21)$$

4.4 ORTHOTROPIC MATERIAL

An orthotropic material has three mechanical properties along three principal axes. To simulate this type of material, the solid in section 4.3 will need to be rotated 180° about x-axis, so the material properties along z-axis can be the same, as shown in Fig. 4.7.

Table 4.2 shows the direction cosines between mutually orthogonal axes before and after rotation:

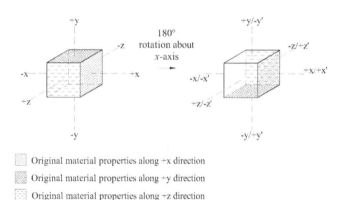

Original material properties along +x direction
Original material properties along +y direction
Original material properties along +z direction

FIGURE 4.7 Solid before and after 180° rotation about x-axis.

TABLE 4.2
Transformation of axis for orthotropic material

	X	y	z
x'	$\cos 0° = 1$	$\cos 90° = 0$	$\cos 90° = 0$
y'	$\cos 90° = 0$	$\cos 180° = -1$	$\cos 90° = 0$
z'	$\cos 90° = 0$	$\cos 90° = 0$	$\cos 180° = -1$

The following is obtained after substitution of the values in Table 4.2 to Eqs. (2.31) and (3.36):

$$
\begin{bmatrix}
\sigma_{x'} & \tau_{x'y'} & \tau_{x'z'} \\
\tau_{y'x'} & \sigma_{y'} & \tau_{y'z'} \\
\tau_{z'x'} & \tau_{z'y'} & \sigma_{z'}
\end{bmatrix}
=
\begin{bmatrix}
1 & 0 & 0 \\
0 & -1 & 0 \\
0 & 0 & -1
\end{bmatrix}
\begin{bmatrix}
\sigma_x & \tau_{xy} & \tau_{xz} \\
\tau_{yx} & \sigma_y & \tau_{yz} \\
\tau_{zx} & \tau_{zy} & \sigma_z
\end{bmatrix}
\begin{bmatrix}
1 & 0 & 0 \\
0 & -1 & 0 \\
0 & 0 & -1
\end{bmatrix}
=
\begin{bmatrix}
\sigma_x & -\tau_{xy} & -\tau_{xz} \\
-\tau_{yx} & \sigma_y & \tau_{yz} \\
-\tau_{zx} & \tau_{zy} & \sigma_z
\end{bmatrix}
$$

$$(4.22)$$

$$
\begin{bmatrix}
\epsilon_{x'} & \tfrac{1}{2}\gamma_{y'x'} & \tfrac{1}{2}\gamma_{z'x'} \\
\tfrac{1}{2}\gamma_{x'y'} & \epsilon_{y'} & \tfrac{1}{2}\gamma_{z'y'} \\
\tfrac{1}{2}\gamma_{x'z'} & \tfrac{1}{2}\gamma_{y'z'} & \epsilon_{z'}
\end{bmatrix}
=
\begin{bmatrix}
1 & 0 & 0 \\
0 & -1 & 0 \\
0 & 0 & -1
\end{bmatrix}
\begin{bmatrix}
\epsilon_x & \tfrac{1}{2}\gamma_{yx} & \tfrac{1}{2}\gamma_{zx} \\
\tfrac{1}{2}\gamma_{xy} & \epsilon_y & \tfrac{1}{2}\gamma_{zy} \\
\tfrac{1}{2}\gamma_{xz} & \tfrac{1}{2}\gamma_{yz} & \epsilon_z
\end{bmatrix}
\begin{bmatrix}
1 & 0 & 0 \\
0 & -1 & 0 \\
0 & 0 & -1
\end{bmatrix}
$$

$$(4.23)$$

$$
=
\begin{bmatrix}
\epsilon_x & -\tfrac{1}{2}\gamma_{xy} & -\tfrac{1}{2}\gamma_{xz} \\
-\tfrac{1}{2}\gamma_{yx} & \epsilon_y & \tfrac{1}{2}\gamma_{yz} \\
-\tfrac{1}{2}\gamma_{zx} & \tfrac{1}{2}\gamma_{zy} & \epsilon_z
\end{bmatrix}
$$

From Eqs. (4.22) and (4.23), the following transformation relationships can be obtained by comparing the components before and after rotation based on their position:

$$
\begin{aligned}
\sigma_{x'} &= \sigma_x \tau_{x'y'} = -\tau_{xy} \\
\sigma_{y'} &= \sigma_y \tau_{x'z'} = -\tau_{xz} \\
\sigma_{z'} &= \sigma_z \tau_{y'z'} = \tau_{yz} \\
\epsilon_{x'} &= \epsilon_x \gamma_{x'y'} = -\gamma_{xy} \\
\epsilon_{y'} &= \epsilon_y \gamma_{x'z'} = -\gamma_{xz} \\
\epsilon_{z'} &= \epsilon_z \gamma_{y'z'} = \gamma_{yz}
\end{aligned}
$$

$$(4.24)$$

σ_x before rotation can be expressed as follows by referring to Eq. (4.21):

$$\sigma_x = D_{11}\epsilon_x + D_{12}\epsilon_y + D_{13}\epsilon_z + D_{14}\gamma_{xy} \qquad (4.25)$$

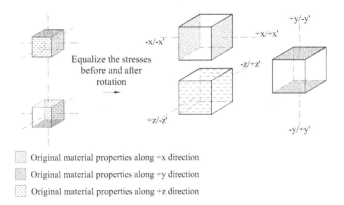

Original material properties along +x direction
Original material properties along +y direction
Original material properties along +z direction

FIGURE 4.8 Transformation from material with symmetrical properties about z axis to orthotropic material.

The expression for $\sigma_{x'}$, which denotes the normal stress in positive x-direction after rotation, can be written based on the general form as shown in Eq. (4.25):

$$\sigma_{x'} = D_{11}\varepsilon_{x'} + D_{12}\varepsilon_{y'} + D_{13}\varepsilon_{z'} + D_{14}\gamma_{x'y'} \tag{4.26}$$

By equating the normal stress in the positive x-direction before and after rotation, we apply the condition where the stresses to be developed in both positive and negative directions along the original global x-axis are the same. This simulates the case where the material properties are the same for the faces normal to x-axis, as shown in Fig. 4.8.

However, Eq. (4.26) needs to be expressed as σ_x in order to do so. By applying the relationships as shown in Eq. (4.24), we can get the following:

$$\sigma_x = D_{11}\varepsilon_x + D_{12}\varepsilon_y + D_{13}\varepsilon_z - D_{14}\gamma_{xy} \tag{4.27}$$

Equating the terms on the right-hand sides in Eqs. (4.25) and (4.27) gives us the following:

$$D_{11}\varepsilon_x + D_{12}\varepsilon_y + D_{13}\varepsilon_z + D_{14}\gamma_{xy} = D_{11}\varepsilon_x + D_{12}\varepsilon_y + D_{13}\varepsilon_z - D_{14}\gamma_{xy}$$

From the comparison of the coefficients of the common terms in equation above, we know that

$$D_{14} = -D_{14}$$

The equation above is valid if and only if D_{14} is zero. The other D components remain the same as they are for a material with symmetrical properties about the z-axis.

σ_y before rotation can be written as below by referring to Eq. (4.21):

$$\sigma_y = D_{21}\varepsilon_x + D_{22}\varepsilon_y + D_{23}\varepsilon_z + D_{24}\gamma_{xy} \qquad (4.28)$$

The expression for $\sigma_{y'}$ can be written based on the general form as shown in Eq. (4.28):

$$\sigma_{y'} = D_{21}\varepsilon_{x'} + D_{22}\varepsilon_{y'} + D_{23}\varepsilon_{z'} + D_{24}\gamma_{x'y'} \qquad (4.29)$$

After applying the relationship in Eq. (4.24) to the equation above, the following expression results:

$$\sigma_y = D_{21}\varepsilon_x + D_{22}\varepsilon_y + D_{23}\varepsilon_z - D_{24}\gamma_{xy} \qquad (4.30)$$

The following result is obtained by equalling the terms on the right-hand side in Eqs. (4.28) and (4.30), then comparing the coefficients of the common terms:

$$D_{24} = -D_{24}$$

The equation above is valid if and only if D_{24} is zero. The other D components remain the same as it is for a material with symmetrical properties about the z-axis.

From the derivation above, we can summarise that the coefficients of σ_x and σ_y (e.g. D_{14}) are zero when stresses before are the same as those after rotation (e.g. $\sigma_x = \sigma_x'$), while the strains associated with those coefficients are equal to their counterparts in opposite sign (e.g. $\gamma_{x'y'} = -\gamma_{xy}$). Therefore, D_{34} and D_{56} are zero because of the following correlations:

$$\sigma_z = D_{31}\varepsilon_x + D_{32}\varepsilon_y + D_{33}\varepsilon_z + D_{34}\gamma_{xy}$$

$$\sigma_z = \sigma_{z'}, \ \gamma_{x'y'} = -\gamma_{xy} \ and \ \gamma_{x'z'} = -\gamma_{xz}$$

$$\tau_{yz} = D_{55}\gamma_{yz} + D_{56}\gamma_{zx}$$

$$\tau_{yz} = \tau_{y'z'}, \ \gamma_{x'y'} = -\gamma_{xy} \ and \ \gamma_{x'z'} = -\gamma_{xz}$$

τ_{xy} before rotation can be expressed as follows by referring to Eq. (4.21):

$$\tau_{xy} = D_{41}\varepsilon_x + D_{42}\varepsilon_y + D_{43}\varepsilon_z + D_{44}\gamma_{xy} \qquad (4.31)$$

$\tau_{xy'}$ can be expressed as below based on the general form as shown in Eq. (4.31):

$$\tau_{x'y'} = D_{41}\varepsilon_{x'} + D_{42}\varepsilon_{y'} + D_{43}\varepsilon_{z'} + D_{44}\gamma_{x'y'} \qquad (4.32)$$

The following is obtained after applying the relationship in Eq. (4.24) to the equation above:

$$- \tau_{xy} = D_{41}\varepsilon_x + D_{42}\varepsilon_y + D_{43}\varepsilon_z - D_{44}\gamma_{xy}$$
$$\tau_{xy} = -D_{41}\varepsilon_x - D_{42}\varepsilon_y - D_{43}\varepsilon_z + D_{44}\gamma_{xy}$$

(4.33)

Equating the terms on the right-hand side in Eqs. (4.31) and (4.33) yields the following:

$$D_{41}\varepsilon_x + D_{42}\varepsilon_y + D_{43}\varepsilon_z + D_{44}\gamma_{xy} = -D_{41}\varepsilon_x - D_{42}\varepsilon_y - D_{43}\varepsilon_z + D_{44}\gamma_{xy}$$

The following expressions are obtained through comparison of the coefficients of common terms in the equation above:

$$D_{41} = -D_{41}$$

$$D_{42} = -D_{42}$$

$$D_{43} = -D_{43}$$

The equations above are valid if and only if D_{41}, D_{42} and D_{43} are zero. The other D components remain the same as they are for material with symmetrical properties about the z-axis.

τ_{xz} before rotation can be expressed as below by referring to Eq. (4.21):

$$\tau_{xz} = D_{65}\gamma_{yz} + D_{66}\gamma_{xz}$$

(4.34)

$\tau_{x'z'}$ can be written in the following form based on the general form as shown in Eq. (4.34):

$$\tau_{x'z'} = D_{65}\gamma_{y'z'} + D_{66}\gamma_{x'z'}$$

(4.35)

With application of relationship in Eq. (4.24) to the equation above, we obtain:

$$- \tau_{xz} = D_{65}\gamma_{yz} - D_{66}\gamma_{zx}$$
$$\tau_{xz} = -D_{65}\gamma_{yz} + D_{66}\gamma_{zx}$$

(4.36)

The following is produced by equating the terms on the right-hand side in Eqs. (4.34) and (4.36), then comparing the coefficients of the common terms:

$$D_{65} = -D_{65}$$

The equation above is valid if and only if D_{65} is zero. The other D components remain the same as they are for a material with properties symmetrical about the z-axis.

With all the derivation above, the expression in Eq. (4.21) is simplified as follows for an orthotropic material:

$$
\begin{bmatrix} \sigma_x \\ \sigma_y \\ \sigma_z \\ \tau_{xy} \\ \tau_{yz} \\ \tau_{zx} \end{bmatrix} = \begin{bmatrix} D_{11} & D_{12} & D_{13} & 0 & 0 & 0 \\ D_{21} & D_{22} & D_{23} & 0 & 0 & 0 \\ D_{31} & D_{32} & D_{33} & 0 & 0 & 0 \\ 0 & 0 & 0 & D_{44} & 0 & 0 \\ 0 & 0 & 0 & 0 & D_{55} & 0 \\ 0 & 0 & 0 & 0 & 0 & D_{66} \end{bmatrix} \begin{bmatrix} \varepsilon_x \\ \varepsilon_y \\ \varepsilon_z \\ \gamma_{xy} \\ \gamma_{yz} \\ \gamma_{zx} \end{bmatrix} \qquad (4.37)
$$

4.5 ORTHOTROPIC MATERIAL WITH SAME PROPERTIES ALONG Y AND Z AXES

The isotropy of orthotropic material can be improved when the properties along two of three mutually orthogonal axes are the same. To simulate this type of material, the solid in section 4.4 will need to be rotated 90° anticlockwise about x-axis, so the material properties along y and z axes can be the same, as shown in Fig. 4.9.

The direction cosines between mutually orthogonal axes before and after rotation are shown in Table 4.3 below:

By substituting the values in Table 4.3 to Eqs. (2.31) and (3.36) produces the following:

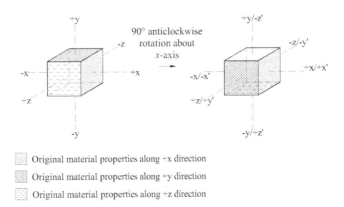

Original material properties along +x direction

Original material properties along +y direction

Original material properties along +z direction

FIGURE 4.9 Solid before and after 90° anticlockwise rotation about x-axis.

TABLE 4.3

Transformation of axis for orthotropic material with same properties along y and z axes

	x	y	z
x'	$\cos 0° = 1$	$\cos 90° = 0$	$\cos 90° = 0$
y'	$\cos 90° = 0$	$\cos 90° = 0$	$\cos 0° = 1$
z'	$\cos 90° = 0$	$\cos 180° = -1$	$\cos 90° = 0$

$$
\begin{bmatrix} \sigma_{x'} \tau_{x'y'} \tau_{x'z'} \\ \tau_{y'x'} \sigma_{y'} \tau_{y'z'} \\ \tau_{z'x'} \tau_{z'y'} \sigma_{z'} \end{bmatrix} = \begin{bmatrix} 1 & 0 & 0 \\ 0 & 0 & 1 \\ 0 & -1 & 0 \end{bmatrix} \begin{bmatrix} \sigma_x \tau_{xy} \tau_{xz} \\ \tau_{yx} \sigma_y \tau_{yz} \\ \tau_{zx} \tau_{zy} \sigma_z \end{bmatrix} \begin{bmatrix} 1 & 0 & 0 \\ 0 & 0 & -1 \\ 0 & 1 & 0 \end{bmatrix} = \begin{bmatrix} \sigma_x & \tau_{xz} & -\tau_{xy} \\ \tau_{zx} & \sigma_z & -\tau_{zy} \\ -\tau_{yx} & -\tau_{yz} & \sigma_y \end{bmatrix}
$$

$$(4.38)$$

$$
\begin{bmatrix} \epsilon_{x'} & \frac{1}{2}\gamma_{y'x'} & \frac{1}{2}\gamma_{z'x'} \\ \frac{1}{2}\gamma_{x'y'} & \epsilon_{y'} & \frac{1}{2}\gamma_{z'y'} \\ \frac{1}{2}\gamma_{x'z'} & \frac{1}{2}\gamma_{y'z'} & \epsilon_{z'} \end{bmatrix} = \begin{bmatrix} 1 & 0 & 0 \\ 0 & 0 & 1 \\ 0 & -1 & 0 \end{bmatrix} \begin{bmatrix} \epsilon_x & \frac{1}{2}\gamma_{yx} & \frac{1}{2}\gamma_{zx} \\ \frac{1}{2}\gamma_{xy} & \epsilon_y & \frac{1}{2}\gamma_{zy} \\ \frac{1}{2}\gamma_{xz} & \frac{1}{2}\gamma_{yz} & \epsilon_z \end{bmatrix} \begin{bmatrix} 1 & 0 & 0 \\ 0 & 0 & -1 \\ 0 & 1 & 0 \end{bmatrix}
$$

$$
= \begin{bmatrix} \epsilon_x & \frac{1}{2}\gamma_{xz} & -\frac{1}{2}\gamma_{xy} \\ \frac{1}{2}\gamma_{zx} & \epsilon_z & -\frac{1}{2}\gamma_{zy} \\ -\frac{1}{2}\gamma_{yx} & -\frac{1}{2}\gamma_{yz} & \epsilon_y \end{bmatrix}
$$

$$(4.39)$$

By comparing the stress and strain components before and after transformation in Eqs. (4.38) and (4.39) respectively, the following relationships result:

$$
\begin{aligned}
\sigma_{x'} &= \sigma_x \tau_{x'y'} = \tau_{xz} \\
\sigma_{y'} &= \sigma_z \tau_{x'z'} = -\tau_{xy} \\
\sigma_{z'} &= \sigma_y \tau_{y'z'} = -\tau_{yz} \\
\epsilon_{x'} &= \epsilon_x \gamma_{x'y'} = \gamma_{xz} \\
\epsilon_{y'} &= \epsilon_z \gamma_{x'z'} = -\gamma_{xy} \\
\epsilon_{z'} &= \epsilon_y \gamma_{y'z'} = -\gamma_{yz}
\end{aligned}
$$

$$(4.40)$$

The expression of σ_x before rotation can be written as follows by referring to Eq. (4.37):

$$
\sigma_x = D_{11}\epsilon_x + D_{12}\epsilon_y + D_{13}\epsilon_z \tag{4.41}
$$

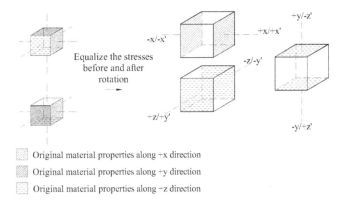

Original material properties along +x direction
Original material properties along +y direction
Original material properties along +z direction

FIGURE 4.10 Transformation from orthotropic material to orthotropic material with symmetrical properties along y and z axes.

The expression for $\sigma_{x'}$, which denotes the normal stress in the positive x-direction after rotation, can be written based on the general form as shown in Eq. (4.41):

$$\sigma_{x'} = D_{11}\varepsilon_{x'} + D_{12}\varepsilon_{y'} + D_{13}\varepsilon_{z'} \tag{4.42}$$

By equating the normal stress in the positive x-direction before and after rotation, we apply the condition where the stresses to be developed in both positive and negative direction along the original global x-axis are the same. This simulates the case where the material properties are the same for the faces normal to x-axis, as shown in Fig. 4.10.

Eq. (4.42) needs to be expressed as σ_x beforehand. By applying the relationships as shown in Eq. (4.40), this can be obtained:

$$\sigma_x = D_{11}\varepsilon_x + D_{12}\varepsilon_z + D_{13}\varepsilon_y \tag{4.43}$$

Equating the terms on the right-hand side in Eqs. (4.41) and (4.43) leads to the following:

$$D_{11}\varepsilon_x + D_{12}\varepsilon_y + D_{13}\varepsilon_z = D_{11}\varepsilon_x + D_{12}\varepsilon_z + D_{13}\varepsilon_y$$

Comparison of the coefficients of the common terms results in the following:

$$D_{12} = D_{13}$$

The expression for σ_y before rotation can be written as follows by referring to Eq. (4.37):

$$\sigma_y = D_{21}\varepsilon_x + D_{22}\varepsilon_y + D_{23}\varepsilon_z \tag{4.44}$$

The general form as shown in Eq. (4.44) can be served as the basis for development of expression for $\sigma_{y'}$:

$$\sigma_{y'} = D_{21}\varepsilon_{x'} + D_{22}\varepsilon_{y'} + D_{23}\varepsilon_{z'}$$

The following is obtained through the application of relationship in Eq. (4.40) to the equation above:

$$\sigma_z = D_{21}\varepsilon_x + D_{22}\varepsilon_z + D_{23}\varepsilon_y \tag{4.45}$$

The expression for σ_y before rotation is necessary to proceed with the derivation. By referring to Eq. (4.37), σ_y before rotation can be expressed as:

$$\sigma_z = D_{31}\varepsilon_x + D_{32}\varepsilon_y + D_{33}\varepsilon_z \tag{4.46}$$

The expression for $\sigma_{z'}$ can be written based on the general form as shown in Eq. (4.46):

$$\sigma_{z'} = D_{31}\varepsilon_{x'} + D_{32}\varepsilon_{y'} + D_{33}\varepsilon_{z'}$$

After applying relationship in Eq. (4.40) to the equation above, we know that:

$$\sigma_y = D_{31}\varepsilon_x + D_{32}\varepsilon_z + D_{33}\varepsilon_y \tag{4.47}$$

Equating the terms on the right-hand side in Eqs. (4.44) and (4.47) and comparing the coefficients of the common terms results in the following:

$$D_{21} = D_{31}$$

$$D_{22} = D_{33}$$

$$D_{23} = D_{32}$$

Same equations will be obtained by repeating the process above for Eqs. (4.45) and (4.46).

The expression for τ_{xy} before rotation is as below by referring to Eq. (4.37):

$$\tau_{xy} = D_{44}\gamma_{xy} \tag{4.48}$$

The expression for $\tau_{x'y'}$ can be written based on the general form as shown in Eq. (4.48):

$$\tau_{x'y'} = D_{44}\gamma_{x'y'}$$

By applying relationship in Eq. (4.40) to the equation above, we get the following equation:

$$\tau_{xz} = D_{44}\gamma_{xz} \tag{4.49}$$

By referring to Eq. (4.37), τ_{xz} before rotation can be expressed as:

$$\tau_{xz} = D_{66}\gamma_{xz} \tag{4.50}$$

By referring to Eq. (4.50), the expression for $\tau_{x'z'}$ is defined as:

$$\tau_{x'z'} = D_{66}\gamma_{x'z'}$$

Applying relationship in Eq. (4.40) to the equation above gives us the follows:

$$\begin{aligned} -\tau_{xy} &= -D_{66}\gamma_{xy} \\ \tau_{xy} &= D_{66}\gamma_{xy} \end{aligned} \tag{4.51}$$

When the terms on the right-hand side in Eqs. (4.48) and (4.51) are equated, comparison of the coefficients for their common terms results in the following:

$$D_{44} = D_{66}$$

Same equations will be obtained if the process above is repeated for Eqs. (4.49) and (4.50).

The expression for τ_{yz} before rotation is as below by referring to Eq. (4.37):

$$\tau_{yz} = D_{55}\gamma_{yz} \tag{4.52}$$

By referring to Eq. (4.52), the expression for $\tau_{y'z'}$ can be written in the following form:

$$\tau_{y'z'} = D_{55}\gamma_{y'z'}$$

After applying the relationship in Eq. (4.40) to equation above, the following is obtained:

$$\begin{aligned} -\tau_{yz} &= -D_{55}\gamma_{yz} \\ \tau_{yz} &= D_{55}\gamma_{yz} \end{aligned} \tag{4.53}$$

By equating the terms on the right-hand side in Eqs. (4.52) and (4.53), then comparing the coefficients of the common terms yields the following:

$$D_{55} = D_{55}$$

With all the derivation above, the expression in Eq. (4.37) is simplified as follows for an orthotropic material with the same properties along x and y axes:

$$
\begin{bmatrix} \sigma_x \\ \sigma_y \\ \sigma_z \\ \tau_{xy} \\ \tau_{yz} \\ \tau_{zx} \end{bmatrix} = \begin{bmatrix} D_{11} & D_{12} & D_{12} & 0 & 0 & 0 \\ D_{21} & D_{22} & D_{23} & 0 & 0 & 0 \\ D_{21} & D_{23} & D_{22} & 0 & 0 & 0 \\ 0 & 0 & 0 & D_{44} & 0 & 0 \\ 0 & 0 & 0 & 0 & D_{55} & 0 \\ 0 & 0 & 0 & 0 & 0 & D_{44} \end{bmatrix} \begin{bmatrix} \varepsilon_x \\ \varepsilon_y \\ \varepsilon_z \\ \gamma_{xy} \\ \gamma_{yz} \\ \gamma_{zx} \end{bmatrix} \tag{4.54}
$$

4.6   HOMOGENEOUS MATERIAL

Homogeneous material is a type of material with the same properties along any mutually orthogonal axes. To simulate this type of material, the solid in section 4.5 will need to be rotated 90° clockwise about the z-axis, so the material properties along x and y axes can be the same. Since the material properties along the z-axis are same as y-axis, the properties along the x-axis will also be similar to those along the z-axis, and thus the simulation matches the characteristics of a homogeneous material, as shown in Fig. 4.11.

Table 4.4 shows the direction cosines between mutually orthogonal axes before and after rotation:

The following equations are obtained after substituting the values in Table 4.4 to Eqs. (2.31) and (3.36):

$$
\begin{bmatrix} \sigma_{x'}\tau_{x'y'}\tau_{x'z'} \\ \tau_{y'x'}\sigma_{y'}\tau_{y'z'} \\ \tau_{z'x'}\tau_{z'y'}\sigma_{z'} \end{bmatrix} = \begin{bmatrix} 0 & 1 & 0 \\ -1 & 0 & 0 \\ 0 & 0 & 1 \end{bmatrix} \begin{bmatrix} \sigma_x\tau_{xy}\tau_{xz} \\ \tau_{yx}\sigma_y\tau_{yz} \\ \tau_{zx}\tau_{zy}\sigma_z \end{bmatrix} \begin{bmatrix} 0 & -1 & 0 \\ 1 & 0 & 0 \\ 0 & 0 & 1 \end{bmatrix} = \begin{bmatrix} \sigma_y & -\tau_{yx} & \tau_{yz} \\ -\tau_{xy} & \sigma_x & -\tau_{xz} \\ \tau_{zy} & -\tau_{zx} & \sigma_z \end{bmatrix}
$$

$$\tag{4.55}$$

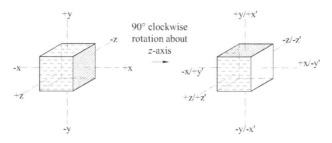

Original material properties along +x direction

Original material properties along +y and +z directions

FIGURE 4.11   Solid before and after 90° clockwise rotation about the z-axis.

TABLE 4.4

Transformation of axis for a homogeneous material

	x	y	z
x'	$\cos 90° = 0$	$\cos 0° = 1$	$\cos 90° = 0$
y'	$\cos 180° = -1$	$\cos 90° = 0$	$\cos 90° = 0$
z'	$\cos 90° = 0$	$\cos 90° = 0$	$\cos 0° = 1$

$$
\begin{bmatrix} \epsilon_{x'} & \frac{1}{2}\gamma_{y'x'} & \frac{1}{2}\gamma_{z'x'} \\ \frac{1}{2}\gamma_{x'y'} & \epsilon_{y'} & \frac{1}{2}\gamma_{z'y'} \\ \frac{1}{2}\gamma_{x'z'} & \frac{1}{2}\gamma_{y'z'} & \epsilon_{z'} \end{bmatrix} = \begin{bmatrix} 0 & 1 & 0 \\ -1 & 0 & 0 \\ 0 & 0 & 1 \end{bmatrix} \begin{bmatrix} \epsilon_{x} & \frac{1}{2}\gamma_{yx} & \frac{1}{2}\gamma_{zx} \\ \frac{1}{2}\gamma_{xy} & \epsilon_{y} & \frac{1}{2}\gamma_{zy} \\ \frac{1}{2}\gamma_{xz} & \frac{1}{2}\gamma_{yz} & \epsilon_{z} \end{bmatrix} \begin{bmatrix} 0 & -1 & 0 \\ 1 & 0 & 0 \\ 0 & 0 & 1 \end{bmatrix}
$$

$$
= \begin{bmatrix} \epsilon_{y} & -\frac{1}{2}\gamma_{yx} & \frac{1}{2}\gamma_{yz} \\ -\frac{1}{2}\gamma_{xy} & \epsilon_{x} & -\frac{1}{2}\gamma_{xz} \\ \frac{1}{2}\gamma_{zy} & -\frac{1}{2}\gamma_{zx} & \epsilon_{z} \end{bmatrix}
$$

(4.56)

With the comparison of stress and strain components before and after rotation in Eqs. (4.55) and (4.56), the following transformation relationships were obtained:

$$
\begin{aligned}
\sigma_{x'} &= \sigma_{y} & \tau_{x'y'} &= -\tau_{xy} \\
\sigma_{y'} &= \sigma_{x} & \tau_{x'z'} &= \tau_{yz} \\
\sigma_{z'} &= \sigma_{z} & \tau_{y'z'} &= -\tau_{xz} \\
\varepsilon_{x'} &= \varepsilon_{y} & \gamma_{x'y'} &= -\gamma_{xy} \\
\varepsilon_{y'} &= \varepsilon_{x} & \gamma_{x'z'} &= \gamma_{yz} \\
\varepsilon_{z'} &= \varepsilon_{z} & \gamma_{y'z'} &= -\gamma_{xz}
\end{aligned}
$$

(4.57)

σ_x before rotation can be expressed as below by referring to Eq. (4.54):

$$
\sigma_x = D_{11}\varepsilon_x + D_{12}\varepsilon_y + D_{12}\varepsilon_z
$$

(4.58)

The expression for $\sigma_{x'}$, which denotes the normal stress in positive x-direction after rotation, can be written based on the general form as shown in Eq. (4.58):

$$
\sigma_{x'} = D_{11}\varepsilon_{x'} + D_{12}\varepsilon_{y'} + D_{12}\varepsilon_{z'}
$$

(4.59)

By equating the normal stress in the positive x-direction before and after rotation, we apply the condition where the stresses to be developed in both the positive and negative directions along the original global x-axis are the same. This simulates the

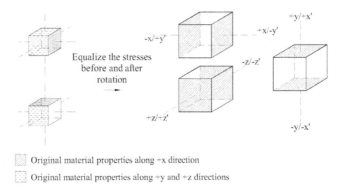

Original material properties along +x direction

Original material properties along +y and +z directions

FIGURE 4.12 Transformation from orthotropic material with symmetrical properties along y and z axes to homogeneous material.

case where the material properties are the same for the faces normal to the x-axis, as shown in Fig. 4.12.

Eq. (4.59) needs to be expressed as σ_x in the first place. This can be obtained by applying the relationships as shown in Eq. (4.57):

$$\sigma_y = D_{11}\varepsilon_y + D_{12}\varepsilon_x + D_{12}\varepsilon_z \tag{4.60}$$

The expression for σ_y before rotation is required for further derivation. It can be expressed as follows by referring to Eq. (4.54):

$$\sigma_y = D_{21}\varepsilon_x + D_{22}\varepsilon_y + D_{23}\varepsilon_z \tag{4.61}$$

The expression for $\sigma_{y'}$ can be written based on the general form as shown in Eq. (4.61):

$$\sigma_{y'} = D_{21}\varepsilon_{x'} + D_{22}\varepsilon_{y'} + D_{23}\varepsilon_{z'}$$

By applying relationship in Eq. (4.57) to the equation above, we can obtain:

$$\sigma_x = D_{21}\varepsilon_y + D_{22}\varepsilon_x + D_{23}\varepsilon_z \tag{4.62}$$

Eqs. (4.58) and (4.62) are comparable as they both express σ_x component. By equalling the terms on the right-hand side in Eqs. (4.58) and (4.62), then comparing the coefficients of the common terms leads to the following:

$$D_{11} = D_{22}$$

$$D_{12} = D_{21} \ and \ D_{12} = D_{23}$$

Same equations will be obtained by repeating the process above for Eqs. (4.60) and (4.61).

σ_z before rotation can be expressed as below by referring to Eq. (4.54):

$$\sigma_z = D_{21}\varepsilon_x + D_{23}\varepsilon_y + D_{22}\varepsilon_z \tag{4.63}$$

The expression for $\sigma_{z'}$ can be written based on the general form as shown in Eq. (4.63):

$$\sigma_{z'} = D_{21}\varepsilon_{x'} + D_{23}\varepsilon_{y'} + D_{22}\varepsilon_{z'}$$

After the relationship in Eq. (4.57) is applied to the equation above, the following is obtained:

$$\sigma_z = D_{21}\varepsilon_y + D_{23}\varepsilon_x + D_{22}\varepsilon_z \tag{4.64}$$

Equalization the terms on the right-hand side in Eqs. (4.63) and (4.64), follows with the comparison of the coefficients of the common terms results in the following:

$$D_{21} = D_{23}$$

$$D_{22} = D_{22}$$

The expression for τ_{xy} before rotation can be derived as below by referring to Eq. (4.54):

$$\tau_{xy} = D_{44}\gamma_{xy} \tag{4.65}$$

The expression for $\tau_{x'y'}$ can be written based on the general form as shown in Eq. (4.65):

$$\tau_{x'y'} = D_{44}\gamma_{x'y'}$$

By taking the relationship as per Eq. (4.57) into account, the following can be obtained from the equation above:

$$- \tau_{xy} = -D_{44}\gamma_{xy}$$
$$\tau_{xy} = D_{44}\gamma_{xy} \tag{4.66}$$

By equating the terms on the right-hand side in Eqs. (4.65) and (4.66), then comparing the coefficients of the common terms gives us the following:

$$D_{44} = D_{44}$$

The expression for τ_{yz} before rotation can be developed by referring to Eq. (4.54):

$$\tau_{yz} = D_{55}\gamma_{yz} \qquad (4.67)$$

Based on the general form as shown in Eq. (4.67), the expression for $\tau_{y'z'}$ can be written as follows:

$$\tau_{y'z'} = D_{55}\gamma_{y'z'}$$

Application of relationship in Eq. (4.57) to the equation above results in the following:

$$\begin{aligned} -\tau_{xz} &= -D_{55}\gamma_{xz} \\ \tau_{xz} &= D_{55}\gamma_{xz} \end{aligned} \qquad (4.68)$$

By referring to Eq. (4.54), τ_{xz} before rotation can be expressed as:

$$\tau_{xz} = D_{44}\gamma_{xz} \qquad (4.69)$$

The expression for $\tau_{x'z'}$ can be written by referring to the general form as shown in Eq. (4.69):

$$\tau_{x'z'} = D_{44}\gamma_{x'z'}$$

By considering the relationship in Eq. (4.57), the equation above gives us:

$$\tau_{yz} = D_{44}\gamma_{yz} \qquad (4.70)$$

After equalizing both Eqs. (4.68) and (4.69), the following is obtained by comparing the coefficients of the common terms:

$$D_{44} = D_{55}$$

Same equations will be obtained by repeat the process above for Eqs. (4.67) and (4.70).

With all the derivation above, the expression in Eq. (4.54) is simplified as follows for a homogeneous material:

$$\begin{bmatrix} \sigma_x \\ \sigma_y \\ \sigma_z \\ \tau_{xy} \\ \tau_{yz} \\ \tau_{zx} \end{bmatrix} = \begin{bmatrix} D_{11} & D_{12} & D_{12} & 0 & 0 & 0 \\ D_{12} & D_{11} & D_{12} & 0 & 0 & 0 \\ D_{12} & D_{12} & D_{11} & 0 & 0 & 0 \\ 0 & 0 & 0 & D_{44} & 0 & 0 \\ 0 & 0 & 0 & 0 & D_{44} & 0 \\ 0 & 0 & 0 & 0 & 0 & D_{44} \end{bmatrix} \begin{bmatrix} \varepsilon_x \\ \varepsilon_y \\ \varepsilon_z \\ \gamma_{xy} \\ \gamma_{yz} \\ \gamma_{zx} \end{bmatrix} \qquad (4.71)$$

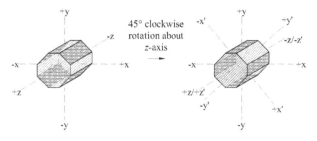

☐ Original material properties along +x, +y and +z directions

☰ Original material properties along directions other than +x, +y and +z directions

FIGURE 4.13 Solid before and after 45° clockwise rotation about z-axis.

TABLE 4.5

Transformation of axis for isotropic material

	x	y	z
x'	$\cos 45° = -\frac{1}{\sqrt{2}}$	$\cos 135° = -\frac{1}{\sqrt{2}}$	$\cos 90° = 0$
y'	$\cos 45° = \frac{1}{\sqrt{2}}$	$\cos 45° = \frac{1}{\sqrt{2}}$	$\cos 90° = 0$
z'	$\cos 90° = 0$	$\cos 90° = 0$	$\cos 0° = 1$

4.7 ISOTROPIC MATERIAL

Isotropic material possesses identical properties along any direction, other than only three mutually orthogonal axes. To simulate this type of material, the solid in section 4.6 will need to be rotated 45° clockwise about the z-axis, so the material properties in any direction can be the same, as shown in Fig. 4.13.

The direction cosines between mutually orthogonal axes before and after rotation are shown in Table 4.5 below:

Substituting the values in Table 4.5 to Eqs. (2.31) and (3.36) yields the following:

$$
\begin{bmatrix}
\sigma_x' \tau_{x'y'} \tau_{x'z'} \\
\tau_{y'x'} \sigma_y' \tau_{y'z'} \\
\tau_{z'x'} \tau_{z'y'} \sigma_z'
\end{bmatrix}
=
\begin{bmatrix}
\frac{1}{\sqrt{2}} & -\frac{1}{\sqrt{2}} & 0 \\
\frac{1}{\sqrt{2}} & \frac{1}{\sqrt{2}} & 0 \\
0 & 0 & 1
\end{bmatrix}
\begin{bmatrix}
\sigma_x \tau_{xy} \tau_{xz} \\
\tau_{yx} \sigma_y \tau_{yz} \\
\tau_{zx} \tau_{zy} \sigma_z
\end{bmatrix}
\begin{bmatrix}
\frac{1}{\sqrt{2}} & \frac{1}{\sqrt{2}} & 0 \\
-\frac{1}{\sqrt{2}} & \frac{1}{\sqrt{2}} & 0 \\
0 & 0 & 1
\end{bmatrix}
$$

$$
=
\begin{bmatrix}
\frac{\sigma_x + \sigma_y - 2\tau_{xy}}{2} & \frac{\sigma_x - \sigma_y}{2} & \frac{\tau_{xz} - \tau_{yz}}{\sqrt{2}} \\
\frac{\sigma_x - \sigma_y}{2} & \frac{\sigma_x + \sigma_y + 2\tau_{xy}}{2} & \frac{\tau_{xz} + \tau_{yz}}{\sqrt{2}} \\
\frac{\tau_{zx} - \tau_{zy}}{\sqrt{2}} & \frac{\tau_{zx} + \tau_{zy}}{\sqrt{2}} & \sigma_z
\end{bmatrix}
\tag{4.72}
$$

$$
\begin{bmatrix}
\epsilon_{x'} & \frac{1}{2}\gamma_{y'x'} & \frac{1}{2}\gamma_{z'x'} \\
\frac{1}{2}\gamma_{x'y'} & \epsilon_{y'} & \frac{1}{2}\gamma_{z'y''} \\
\frac{1}{2}\gamma_{x'z'} & \frac{1}{2}\gamma_{y'z'} & \epsilon_{z'}
\end{bmatrix}
=
\begin{bmatrix}
\frac{1}{\sqrt{2}} & -\frac{1}{\sqrt{2}} & 0 \\
\frac{1}{\sqrt{2}} & \frac{1}{\sqrt{2}} & 0 \\
0 & 0 & 1
\end{bmatrix}
\begin{bmatrix}
\epsilon_x & \frac{1}{2}\gamma_{yx} & \frac{1}{2}\gamma_{zx} \\
\frac{1}{2}\gamma_{xy} & \epsilon_y & \frac{1}{2}\gamma_{zy} \\
\frac{1}{2}\gamma_{xz} & \frac{1}{2}\gamma_{yz} & \epsilon_z
\end{bmatrix}
\begin{bmatrix}
\frac{1}{\sqrt{2}} & \frac{1}{\sqrt{2}} & 0 \\
-\frac{1}{\sqrt{2}} & \frac{1}{\sqrt{2}} & 0 \\
0 & 0 & 1
\end{bmatrix}
$$

$$
=
\begin{bmatrix}
\frac{\epsilon_x + \epsilon_y - \gamma_{xy}}{2} & \frac{\epsilon_x - \epsilon_y}{2} & \frac{\gamma_{xz} - \gamma_{yz}}{2\sqrt{2}} \\
\frac{\epsilon_x - \epsilon_y}{2} & \frac{\epsilon_x + \epsilon_y + \gamma_{xy}}{2} & \frac{\gamma_{xz} + \gamma_{yz}}{2\sqrt{2}} \\
\frac{\gamma_{zx} - \gamma_{zy}}{2\sqrt{2}} & \frac{\gamma_{zx} + \gamma_{zy}}{2\sqrt{2}} & \epsilon_z
\end{bmatrix}
$$

$$\tag{4.73}$$

The following transformation relationships can be obtained by comparing the stress and strain components before and after rotation based in Eqs. (4.72) and (4.73), respectively:

$$
\begin{aligned}
\sigma_{x'} &= \frac{\sigma_x + \sigma_y - 2\tau_{xy}}{2} & \tau_{x'y'} &= \frac{\sigma_x - \sigma_y}{2} \\
\sigma_{y'} &= \frac{\sigma_x + \sigma_y + 2\tau_{xy}}{2} & \tau_{x'z'} &= \frac{\tau_{xz} - \tau_{yz}}{\sqrt{2}} \\
\sigma_{z'} &= \sigma_z & \tau_{y'z'} &= \frac{\tau_{xz} + \tau_{yz}}{\sqrt{2}} \\
\epsilon_{x'} &= \frac{\epsilon_x + \epsilon_y - \gamma_{xy}}{2} & \frac{1}{2}\gamma_{x'y'} &= \frac{\epsilon_x - \epsilon_y}{2}, & \gamma_{x'y'} &= \epsilon_x - \epsilon_y \\
\epsilon_{y'} &= \frac{\epsilon_x + \epsilon_y + \gamma_{xy}}{2} & \frac{1}{2}\gamma_{x'z'} &= \frac{\gamma_{xz} - \gamma_{yz}}{2\sqrt{2}}, & \gamma_{x'z'} &= \frac{\gamma_{xz} - \gamma_{yz}}{\sqrt{2}} \\
\epsilon_{z'} &= \epsilon_z & \frac{1}{2}\gamma_{y'z'} &= \frac{\gamma_{xz} + \gamma_{yz}}{2\sqrt{2}}, & \gamma_{y'z'} &= \frac{\gamma_{xz} + \gamma_{yz}}{\sqrt{2}}
\end{aligned}
$$

$$\tag{4.74}$$

By referring to Eq. (4.71), σ_x, σ_y and τ_{xy} before rotation can be expressed as follows:

$$
\begin{aligned}
\sigma_x &= D_{11}\epsilon_x + D_{12}\epsilon_y + D_{12}\epsilon_z \\
\sigma_y &= D_{12}\epsilon_x + D_{11}\epsilon_y + D_{12}\epsilon_z \\
\tau_{xy} &= D_{44}\gamma_{xy}
\end{aligned}
$$

$$\tag{4.75}$$

The expression for $\sigma_{x'}$, which denotes the normal stress in positive x-direction after rotation, can be written based on the general form as shown in Eq. (4.75):

$$
\sigma_{x'} = D_{11}\epsilon_{x'} + D_{12}\epsilon_{y'} + D_{12}\epsilon_{z'}
$$

$$\tag{4.76}$$

By equating the normal stress in the positive x-direction before and after rotation, we apply the condition where the stresses to be developed in both positive and negative direction along the original global x-axis are the same. This simulates the case where the material properties are the same for the faces normal to x-axis, as shown in Fig. 4.14.

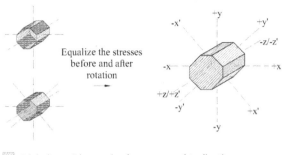

Original material properties along +x, +y and +z directions

Original material properties along directions other than +x, +y and +z directions

FIGURE 4.14 Transformation from homogeneous material to isotropic material.

Eq. (4.76) needs to be expressed as σ_x before proceeding with further derivation. This can be achieved by applying the relationships as shown in Eq. (4.74):

$$\frac{\sigma_x + \sigma_y - 2\tau_{xy}}{2} = D_{11}\left(\frac{\varepsilon_x + \varepsilon_y - \gamma_{xy}}{2}\right) + D_{12}\left(\frac{\varepsilon_x + \varepsilon_y + \gamma_{xy}}{2}\right) + D_{12}\varepsilon_z$$

Multiplying the terms on both sides in the equation above by 2 yields the following:

$$\sigma_x + \sigma_y - 2\tau_{xy} = D_{11}(\varepsilon_x + \varepsilon_y - \gamma_{xy}) + D_{12}(\varepsilon_x + \varepsilon_y + \gamma_{xy}) + 2D_{12}\varepsilon_z$$

Applying the relationship in Eq. (4.75) to the equation above results in the following:

$$\left(D_{11}\varepsilon_x + D_{12}\varepsilon_y + D_{12}\varepsilon_z\right) + \left(D_{12}\varepsilon_x + D_{11}\varepsilon_y + D_{12}\varepsilon_z\right) - 2(D_{44}\gamma_{xy})$$
$$= D_{11}(\varepsilon_x + \varepsilon_y - \gamma_{xy}) + D_{12}(\varepsilon_x + \varepsilon_y + \gamma_{xy}) + 2D_{12}\varepsilon_z$$

By rearranging the terms in the equation above we can get:

$$\varepsilon_x(D_{11} + D_{12}) + \varepsilon_y(D_{11} + D_{12}) + 2D_{12}\varepsilon_z - 2D_{44}\gamma_{xy}$$
$$= \varepsilon_x(D_{11} + D_{12}) + \varepsilon_y(D_{11} + D_{12}) - \gamma_{xy}(D_{11} - D_{12}) + 2D_{12}\varepsilon_z$$

Then, comparing the coefficients of the common terms in the equation above, the following is obtained:

$$-2D_{44} = -(D_{11} - D_{12})$$
$$2D_{44} = D_{11} - D_{12} \tag{4.77}$$

Rearranging Eq. (4.77) yields the following equation:

$$D_{11} = D_{12} + 2D_{44}$$

By introducing Lame Constants, λ and μ to substitute D_{12} and D_{44} in the equation above, the equation is transformed to the following form:

$$D_{11} = \lambda + 2\mu \tag{4.78}$$

Also, by applying the relationship in Eqs. (4.4) and (4.75), it can be concluded that:

$$D_{44} = G$$

Therefore, Eq. (4.78) can be written as follows:

$$D_{11} = \lambda + 2G$$

For an isotropic material, its D matrix is the same as that for a homogeneous material Eq. (4.71). However, the unknowns, i.e. D_{11}, D_{12} and D_{44}, are related as shown in the equation above. In this case, only two parameters are required to determine the D matrix:

$$\begin{bmatrix} \sigma_x \\ \sigma_y \\ \sigma_z \\ \tau_{xy} \\ \tau_{yz} \\ \tau_{zx} \end{bmatrix} = \begin{bmatrix} \lambda + 2G & \lambda & \lambda & 0 & 0 & 0 \\ \lambda & \lambda + 2G & \lambda & 0 & 0 & 0 \\ \lambda & \lambda & \lambda + 2G & 0 & 0 & 0 \\ 0 & 0 & 0 & G & 0 & 0 \\ 0 & 0 & 0 & 0 & G & 0 \\ 0 & 0 & 0 & 0 & 0 & G \end{bmatrix} \begin{bmatrix} \varepsilon_x \\ \varepsilon_y \\ \varepsilon_z \\ \gamma_{xy} \\ \gamma_{yz} \\ \gamma_{zx} \end{bmatrix} \tag{4.79}$$

Strain, say ε_x, is not only an effect of stress, σ_x. For every unit of ε_y and ε_z due to σ_y and σ_z, a certain amount of ε_x will be induced. Such a phenomenon is described using Poisson's ratio (Eq. (4.3)):

$$\nu = -\frac{\varepsilon_x}{\varepsilon_y} \quad and \quad \nu = -\frac{\varepsilon_x}{\varepsilon_z}$$

Poisson's ratio is the same for every direction for isotropic material. Therefore, by expressing ε_x in terms of other parameters, the following equations result:

$$\varepsilon_x = -\nu\varepsilon_y \quad and \quad \varepsilon_x = -\nu\varepsilon_z$$

By applying the relationship in Eq. (4.1), ε_x can be expressed in terms of σ_y and σ_z:

$$\varepsilon_x = -\frac{\nu}{E}\sigma_y \quad and \quad \varepsilon_x = -\frac{\nu}{E}\sigma_z$$

Total strain is the summation of the effect of stresses in all directions. For x-direction, the total strain is:

$$\varepsilon_x = \frac{\sigma_x}{E} - \frac{\nu}{E}\sigma_y - \frac{\nu}{E}\sigma_z$$

After simplification, the equation above is transformed to the following form:

$$\varepsilon_x = \frac{1}{E}\left[\sigma_x - \nu(\sigma_y + \sigma_z)\right] \tag{4.80}$$

Similarly, the expression for strain in y and z directions can be written as follows:

$$\varepsilon_y = \frac{1}{E}\left[\sigma_y - \nu(\sigma_x + \sigma_z)\right] \tag{4.81}$$

$$\varepsilon_z = \frac{1}{E}\left[\sigma_z - \nu(\sigma_x + \sigma_y)\right] \tag{4.82}$$

Let e = $\varepsilon_x + \varepsilon_y + \varepsilon_z$, by substituting Eqs. (4.80), (4.81) and (4.82) yields the following:

$$e = \left[\frac{\sigma_x}{E} - \frac{\nu}{E}(\sigma_y + \sigma_z)\right] + \left[\frac{\sigma_y}{E} - \frac{\nu}{E}(\sigma_x + \sigma_z)\right] + \left[\frac{\sigma_z}{E} - \frac{\nu}{E}(\sigma_x + \sigma_y)\right]$$

With expansion the equation above becomes:

$$e = \frac{\sigma_x}{E} - \frac{\nu}{E}\sigma_y - \frac{\nu}{E}\sigma_z + \frac{\sigma_y}{E} - \frac{\nu}{E}\sigma_x - \frac{\nu}{E}\sigma_z + \frac{\sigma_z}{E} - \frac{\nu}{E}\sigma_x - \frac{\nu}{E}\sigma_y$$

By rearranging the equation above we get this:

$$e = \left(\frac{\sigma_x}{E} - \frac{2\nu}{E}\sigma_x\right) + \left(\frac{\sigma_y}{E} - \frac{2\nu}{E}\sigma_y\right) + \left(\frac{\sigma_z}{E} - \frac{2\nu}{E}\sigma_z\right)$$

The equation above can be simplified, and it leads to this:

$$e = \sigma_x\left(\frac{1-2\nu}{E}\right) + \sigma_y\left(\frac{1-2\nu}{E}\right) + \sigma_z\left(\frac{1-2\nu}{E}\right)$$

$$e = \frac{1-2\nu}{E}(\sigma_x + \sigma_y + \sigma_z) \tag{4.83}$$

$$\sigma_x + \sigma_y + \sigma_z = \frac{Ee}{1-2\nu}$$

If hydrostatic stress, σ_m applied on the element, then the equation above will be transformed into the following form by substituting Eq. (2.45):

$$3\sigma_m = \frac{Ee}{1 - 2\nu}$$

$$\sigma_m = \frac{Ee}{3(1 - 2\nu)}$$

By comparing the equation above with general stress–strain relation $\sigma = E\varepsilon$, we can define that:

$$Bulk\ \ modulus,\ \ K = \frac{E}{3(1 - 2\nu)}$$

By substituting Eq. (4.83) into Eq. (4.80) we can obtain the follows:

$$\varepsilon_x = \frac{1}{E}\left[\sigma_x - \nu\left(\frac{Ee}{1 - 2\nu} - \sigma_x\right)\right]$$

Expansion of the equation above results in the following equation:

$$E\varepsilon_x = \sigma_x - \frac{\nu Ee}{1 - 2\nu} + \nu\sigma_x$$

$$E\varepsilon_x = \sigma_x(1 + \nu) - \frac{\nu Ee}{1 - 2\nu}$$

Moving the term σ_x to the other side, we get:

$$\sigma_x(1 + \nu) = E\varepsilon_x + \frac{\nu Ee}{1 - 2\nu}$$

By expressing σ_x in terms of other parameters, the following equation is obtained:

$$\sigma_x = \frac{E\varepsilon_x}{1 + \nu} + \frac{\nu Ee}{(1 - 2\nu)(1 + \nu)}$$

Reducing the expression above to only one fractional term yields the following:

$$\sigma_x = \frac{E(1 - 2\nu)\varepsilon_x + \nu Ee}{(1 - 2\nu)(1 + \nu)}$$

After expansion, the equation above becomes:

$$\sigma_x = \frac{E\varepsilon_x - 2E\nu\varepsilon_x + \nu Ee}{(1 - 2\nu)(1 + \nu)}$$

Factorising the numerator with common term E results in the following:

$$\sigma_x = \frac{E(\varepsilon_x - 2\nu\varepsilon_x + \nu e)}{(1 - 2\nu)(1 + \nu)}$$

The equation above can be simplified to the following form:

$$\sigma_x = \frac{E}{(1 - 2\nu)(1 + \nu)}[\varepsilon_x(1 - 2\nu) + \nu e] \tag{4.84}$$

From Eq. (4.79), σ_x can be expressed as:

$$\sigma_x = \varepsilon_x(\lambda + 2G) + \lambda\varepsilon_y + \lambda\varepsilon_z$$

Rewriting the equation above yields the following:

$$\sigma_x = \lambda\varepsilon_x + \lambda\varepsilon_y + \lambda\varepsilon_z + 2G\varepsilon_x$$

By substituting $e = \varepsilon_x + \varepsilon_y + \varepsilon_z$ into the equation above, we obtain the following equation:

$$\sigma_x = \lambda e + 2G\varepsilon_x \tag{4.85}$$

Equating Eqs. (4.84) and (4.85) leads to the following:

$$\frac{E}{(1 - 2\nu)(1 + \nu)}[\varepsilon_x(1 - 2\nu) + \nu e] = \lambda e + 2G\varepsilon_x$$

Expanding the left-hand side of the equation above gives us the follows:

$$\frac{E(1 - 2\nu)}{(1 - 2\nu)(1 + \nu)}\varepsilon_x + \frac{E\nu}{(1 - 2\nu)(1 + \nu)}e = \lambda e + 2G\varepsilon_x$$

$$\frac{E}{(1 + \nu)}\varepsilon_x + \frac{E\nu}{(1 - 2\nu)(1 + \nu)}e = \lambda e + 2G\varepsilon_x$$

By comparing the coefficients of ε_x and e. equation above, we can say that:

$$\lambda = \frac{E\nu}{(1 - 2\nu)(1 + \nu)} \tag{4.86}$$

$$2G = \frac{E}{(1+\nu)}$$

$$G = \frac{E}{2(1+\nu)} \tag{4.87}$$

From Eq. (4.86), the following equation is obtained by expressing E in terms of other parameters:

$$E = \frac{\lambda(1 - 2\nu)(1 + \nu)}{\nu} \tag{4.88}$$

Similarly, Eq. (4.87) can be transformed as follows by putting only E on the left-hand side of the equation:

$$E = 2G(1 + \nu) \tag{4.89}$$

Eqs. (4.88) and (4.89) are now equal and the following is obtained:

$$\frac{\lambda(1 - 2\nu)(1 + \nu)}{\nu} = 2G(1 + \nu)$$

From the equation above, by expressing λ in terms of G and ν yields the following:

$$\lambda = \frac{2G\nu}{(1 - 2\nu)} \tag{4.90}$$

On the other hand, the equation above can be transformed by expressing G in terms of λ and ν:

$$G = \frac{\lambda(1 - 2\nu)}{2\nu}$$

Expressing the equation above with ν in terms of λ and G yields the following:

$$2\nu G = \lambda - 2\lambda\nu$$
$$2\nu G + 2\lambda\nu = \lambda$$
$$2(G + \lambda)\nu = \lambda$$
$$\nu = \frac{\lambda}{2(G + \lambda)} \tag{4.91}$$

Expansion of Eq. (4.89) results in the following:

$$E = 2G + 2G\nu$$

$$\nu = \frac{E - 2G}{2G}$$

By substituting Eq. (4.91) into the equation above gives us the following equation:

$$\frac{\lambda}{2(G + \lambda)} = \frac{E - 2G}{2G}$$

Expressing equation above with E in terms of λ and G and we get this:

$$E = \frac{2G\lambda}{2(G + \lambda)} + 2G$$

$$E = \frac{2G\lambda + 2G(2G + 2\lambda)}{2(G + \lambda)}$$

$$E = \frac{2G\lambda + 4G^2 + 4G\lambda}{2(G + \lambda)}$$

$$E = \frac{6G\lambda + 4G^2}{2(G + \lambda)}$$

The equation above can be factorised with G and as a result:

$$E = \frac{G(4G + 6\lambda)}{2(G + \lambda)}$$

By dividing the numerator and denominator by 2, the equation above can be simplified to the following form:

$$E = \frac{G(2G + 3\lambda)}{G + \lambda}$$

4.8 PLANE STRESS AND PLANE STRAIN FOR AN ISOTROPIC MATERIAL

Under plane stress condition, $\sigma_z = \tau_{yz} = \tau_{xz} = 0$. Eq. (4.80) will be written as follows:

$$\varepsilon_x = \frac{1}{E}(\sigma_x - \nu\sigma_y)$$

Express the equation above with σ_y in terms of other parameters results in the equation below:

$$- \nu \sigma_y = E \varepsilon_x - \sigma_x$$

$$\sigma_y = \frac{1}{\nu}(\sigma_x - E \varepsilon_x)$$

Substituting the equation above and $\sigma_z = 0$ into Eq. (4.83) yields the following:

$$\varepsilon_y = \frac{1}{E} \left[\frac{1}{\nu}(\sigma_x - E \varepsilon_x) - \nu(\sigma_x) \right]$$

The expansion of the equation above leads to the following:

$$E \varepsilon_y = \frac{1}{\nu} \sigma_x - \frac{E}{\nu} \varepsilon_x - \nu \sigma_x$$

By rearranging the equation above, we get:

$$\frac{1}{\nu} \sigma_x - \nu \sigma_x = \frac{E}{\nu} \varepsilon_x + E \varepsilon_y$$

The equation above can be simplified, and the following is obtained after simplification:

$$\left(\frac{1}{\nu} - \nu \right) \sigma_x = E \left(\frac{\varepsilon_x}{\nu} + \varepsilon_y \right)$$

$$\left(\frac{1 - \nu^2}{\nu} \right) \sigma_x = E \left(\frac{\varepsilon_x}{\nu} + \varepsilon_y \right)$$

Expressing the equation above with σ_x in terms of other parameters yields the following:

$$\sigma_x = \frac{E \nu}{1 - \nu^2} \left(\frac{\varepsilon_x}{\nu} + \varepsilon_y \right)$$

Expanding the equation above leads to the equation below:

$$\sigma_x = \frac{E}{1 - \nu^2} \varepsilon_x + \frac{E \nu}{1 - \nu^2} \varepsilon_y \qquad (4.92)$$

A similar expression can be written for σ_y:

$$\sigma_y = \frac{E\nu}{1 - \nu^2}\varepsilon_x + \frac{E}{1 - \nu^2}\varepsilon_y \qquad (4.93)$$

For plane stress condition, D matrix in Eq. (4.79) will be reduced into a 3×3 matrix. By substituting Eqs. (4.87), (4.92) and (4.93) into the reduced D matrix:

$$\begin{bmatrix} \sigma_x \\ \sigma_y \\ \tau_{xy} \end{bmatrix} = \begin{bmatrix} \frac{E}{1-\nu^2} & \frac{E\nu}{1-\nu^2} & 0 \\ \frac{E\nu}{1-\nu^2} & \frac{E}{1-\nu^2} & 0 \\ 0 & 0 & \frac{E}{2(1+\nu)} \end{bmatrix} \begin{bmatrix} \varepsilon_x \\ \varepsilon_y \\ \gamma_{xy} \end{bmatrix}$$

Factorising the terms in D matrix above with $\frac{E}{1-\nu^2}$ $\left(\text{also expressed as} \frac{E}{(1+\nu)(1-\nu)}\right)$ yields the following:

$$\begin{bmatrix} \sigma_x \\ \sigma_y \\ \tau_{xy} \end{bmatrix} = \frac{E}{1 - \nu^2} \begin{bmatrix} 1 & \nu & 0 \\ \nu & 1 & 0 \\ 0 & 0 & \frac{1-\nu}{2} \end{bmatrix} \begin{bmatrix} \varepsilon_x \\ \varepsilon_y \\ \gamma_{xy} \end{bmatrix} \qquad (4.94)$$

Under plane strain condition, $\varepsilon_z = \gamma_{yz} = \gamma_{xz} = 0$, and $\sigma_z \neq 0$. Eq. (4.82) will be written as follows:

$$0 = \frac{1}{E}[\sigma_z - \nu(\sigma_x + \sigma_y)]$$

Expressing the equation above with σ_z in terms of other parameters leads to the following equation:

$$\sigma_z = \nu(\sigma_x + \sigma_y) \qquad (4.95)$$

Substituting Eq. (4.95) into Eq. (4.80) results in the following:

$$\varepsilon_x = \frac{1}{E}[\sigma_x - \nu\left(\sigma_y + \nu(\sigma_x + \sigma_y)\right)]$$

The following is obtained by expanding the equation above:

$$E\varepsilon_x = \sigma_x - \nu(\sigma_y + \nu\sigma_x + \nu\sigma_y)$$

$$E\varepsilon_x = \sigma_x - \nu\sigma_y - \nu^2\sigma_x - \nu^2\sigma_y$$

Rearranging and simplifying the equation above yields the following:

$$E\varepsilon_x = \sigma_x - \nu^2\sigma_x - \nu\sigma_y - \nu^2\sigma_y$$

$$\varepsilon_x = \frac{(1-\nu^2)}{E}\sigma_x - \frac{(\nu+\nu^2)}{E}\sigma_y \tag{4.96}$$

The strain ε_y can be written as below in a similar fashion as Eq. (4.96).

$$\varepsilon_y = -\frac{(\nu+\nu^2)}{E}\sigma_x + \frac{(1-\nu^2)}{E}\sigma_y \tag{4.97}$$

By expressing Eq. (4.4) with $G = \frac{E}{2(1+\nu)}$, Eqs. (4.96) and (4.97) in matrix form yields the follows:

$$
\begin{bmatrix} \varepsilon_x \\ \varepsilon_y \\ \gamma_{xy} \end{bmatrix} =
\begin{bmatrix} \frac{1-\nu^2}{E} & \frac{-(\nu+\nu^2)}{E} & 0 \\ \frac{-(\nu+\nu^2)}{E} & \frac{1-\nu^2}{E} & 0 \\ 0 & 0 & \frac{2(1+\nu)}{E} \end{bmatrix}
\begin{bmatrix} \sigma_x \\ \sigma_y \\ \tau_{xy} \end{bmatrix}
$$

Let X matrix be $\begin{bmatrix} \frac{1-\nu^2}{E} & \frac{-(\nu+\nu^2)}{E} & 0 \\ \frac{-(\nu+\nu^2)}{E} & \frac{1-\nu^2}{E} & 0 \\ 0 & 0 & \frac{2(1+\nu)}{E} \end{bmatrix}$, D matrix can be determined by finding

the inverse matrix of X:

$$
[X]^{-1}\begin{bmatrix} \varepsilon_x \\ \varepsilon_y \\ \gamma_{xy} \end{bmatrix} = [X]^{-1}[X]\begin{bmatrix} \sigma_x \\ \sigma_y \\ \tau_{xy} \end{bmatrix}
$$

$$
\begin{bmatrix} \sigma_x \\ \sigma_y \\ \tau_{xy} \end{bmatrix} = [X]^{-1}\begin{bmatrix} \varepsilon_x \\ \varepsilon_y \\ \gamma_{xy} \end{bmatrix} = [D]\begin{bmatrix} \varepsilon_x \\ \varepsilon_y \\ \gamma_{xy} \end{bmatrix}
$$

The determinant of matrix X defined as follows:

$$|X| = \frac{1-\nu^2}{E}\left[\frac{1-\nu^2}{E} \times \frac{2(1+\nu)}{E}\right] - \frac{-(\nu+\nu^2)}{E}\left[\frac{-(\nu+\nu^2)}{E} \times \frac{2(1+\nu)}{E}\right]$$

$$|X| = \frac{2(1+\nu)(1-\nu^2)^2}{E^3} - \frac{2(1+\nu)(\nu+\nu^2)^2}{E^3}$$

$$|X| = \frac{2(1 + v)[(1 - v^2)^2 - (v + v^2)^2]}{E^3}$$

The following is obtained from the expansion of equation above:

$$|X| = \frac{2(1 + v)[1 - 2v^2 + v^4 - (v^2 + 2v^3 + v^4)]}{E^3}$$

$$|X| = \frac{2(1 + v)(1 - 3v^2 - 2v^3)}{E^3}$$

$(1 - 3v^2 - 2v^3)$ can be simplified into $-(2v - 1)(1 + v)^2$. By substituting this result into equation above gives us the following:

$$|X| = -\frac{2(2v - 1)(1 + v)^3}{E^3}$$

Since the matrix $[X]$ is symmetrical, $[X]^T = [X]$. The adjoint for matrix $[X]$ is defined as follows:

$$adj(X) = \begin{bmatrix} \frac{1-v^2}{E} \times \frac{2(1+v)}{E} & \frac{(v+v^2)}{E} \times \frac{2(1+v)}{E} & 0 \\ \frac{(v+v^2)}{E} \times \frac{2(1+v)}{E} & \frac{1-v^2}{E} \times \frac{2(1+v)}{E} & 0 \\ 0 & 0 & \frac{1-v^2}{E} \times \frac{1-v^2}{E} - \frac{-(v+v^2)}{E} \times \frac{-(v+v^2)}{E} \end{bmatrix}$$

Simplify the adjoint yields the following form:

$$adj(X) = \begin{bmatrix} \frac{2(1+v)(1-v^2)}{E^2} & \frac{2(1+v)(v+v^2)}{E^2} & 0 \\ \frac{2(1+v)(v+v^2)}{E^2} & \frac{2(1+v)(1-v^2)}{E^2} & 0 \\ 0 & 0 & \frac{-(2v-1)(1+v)^2}{E^2} \end{bmatrix}$$

$$adj(X) = \begin{bmatrix} \frac{2(1+v)(1-v^2)}{E^2} & \frac{2(1+v)(v+v^2)}{E^2} & 0 \\ \frac{2(1+v)(v+v^2)}{E^2} & \frac{2(1+v)(1-v^2)}{E^2} & 0 \\ 0 & 0 & \frac{-(2v-1)(1+v)^2}{E^2} \end{bmatrix}$$

D matrix, which is the inverse of matrix X defined as:

$$D = \frac{1}{|X|}adj\,(X) = -\frac{E^3}{2(2v-1)(1+v)^3}\begin{bmatrix} \frac{2(1+v)(1-v^2)}{E^2} & \frac{2(1+v)(v+v^2)}{E^2} & 0 \\ \frac{2(1+v)(v+v^2)}{E^2} & \frac{2(1+v)(1-v^2)}{E^2} & 0 \\ 0 & 0 & \frac{-(2v-1)(1+v)^2}{E^2} \end{bmatrix}$$

The following is obtained by factorizing the terms in matrix above with $\frac{E^2}{2(1+v)}$:

$$D = -\frac{E}{(2v-1)(1+v)^2}\begin{bmatrix} (1-v^2) & v(1+v) & 0 \\ v(1+v) & (1-v^2) & 0 \\ 0 & 0 & \frac{-(2v-1)(1+v)}{2} \end{bmatrix}$$

Factorise the terms in the matrix above with the also expressed as $(1-v)(1+v)$ gives us the follows:

$$D = -\frac{E(1-v)}{(2v-1)(1+v)}\begin{bmatrix} 1 & \frac{v}{1-v} & 0 \\ \frac{v}{1-v} & 1 & 0 \\ 0 & 0 & \frac{-(2v-1)}{2(1-v)} \end{bmatrix}$$

$$D = \frac{E(1-v)}{(1-2v)(1+v)}\begin{bmatrix} 1 & \frac{v}{1-v} & 0 \\ \frac{v}{1-v} & 1 & 0 \\ 0 & 0 & \frac{1-2v}{2(1-v)} \end{bmatrix} \qquad (4.98)$$

For plane strain condition, D matrix in Eq. (4.79) will be reduced into a 3×3 matrix. The remaining stress and strain components are related through the D matrix as shown in Eq. (4.98):

$$\begin{bmatrix} \sigma_x \\ \sigma_y \\ \tau_{xy} \end{bmatrix} = \frac{E(1-v)}{(1-2v)(1+v)}\begin{bmatrix} 1 & \frac{v}{1-v} & 0 \\ \frac{v}{1-v} & 1 & 0 \\ 0 & 0 & \frac{1-2v}{2(1-v)} \end{bmatrix}\begin{bmatrix} \varepsilon_x \\ \varepsilon_y \\ \gamma_{xy} \end{bmatrix} \qquad (4.99)$$

5 Solutions for Elasticity

5.1 INTRODUCTION

Most bodies have linear elastic behaviour, or so they are assumed. Using modulus of elasticity of the material and limiting strain for structural element, the engineer can easily determine the limit stress that can be developed in that element. With this, an engineer can determine the element size for any design load.

Concrete is a brittle material. If it exhibits strain beyond its elastic limit, the bond that holds its aggregates together will be permanently destroyed and only the reinforcing steel remains as a functioning component in the structural element. As a result, the occupant of the building will be at risk. Therefore, engineers design a concrete structure in such a way that ensures the material will not be loaded until inelastic behaviour is induced. Design standards such as Eurocode 2 specifies the design concrete strength and its safety factor based on this philosophy.

In engineering, stress and displacement in a body are often concerned. For elasticity, these parameters can be determined with boundary conditions is identified. Two approaches are commonly used, depends on the objectives of the analysis. Navier equation is a displacement-based approach that determines the displacement over a body given its boundary conditions. Beltrami–Michell stress compatibility equation, on the other hand, is a stress-based approach that aims to determine stress over a body is given the compatibility and boundary conditions to be fulfilled. To derive both methods, stress–displacement relationship, which will be discussed in Section 5.2 is essential.

5.2 STRESS–DISPLACEMENT RELATIONSHIP

For isotropic material, the following stress–strain relationship can be derived from Eq. (4.79):

$$
\begin{aligned}
\sigma_x &= (\lambda + 2G)\varepsilon_x + \lambda(\varepsilon_y + \varepsilon_z) \\
\sigma_y &= (\lambda + 2G)\varepsilon_y + \lambda(\varepsilon_x + \varepsilon_z) \\
\sigma_z &= (\lambda + 2G)\varepsilon_z + \lambda(\varepsilon_x + \varepsilon_y) \\
\tau_{xy} &= G\gamma_{xy} \\
\tau_{yz} &= G\gamma_{yz} \\
\tau_{xz} &= G\gamma_{xz}
\end{aligned}
\tag{5.1}
$$

By applying the relationships in Eqs. (3.2), (3.3), (3.4) to corresponding terms in stress–strain relationships for σ_x as shown in Eq. (5.1) results in the follows:

$$\sigma_x = (\lambda + 2G)\frac{\partial u}{\partial x} + \lambda\left(\frac{\partial v}{\partial y} + \frac{\partial w}{\partial z}\right)$$

Equation below is obtained by expanding equation above:

$$\sigma_x = \lambda\frac{\partial u}{\partial x} + 2G\frac{\partial u}{\partial x} + \lambda\frac{\partial v}{\partial y} + \lambda\frac{\partial w}{\partial z}$$

After rearrangement, the equation above is transformed to the following:

$$\sigma_x = \lambda\left(\frac{\partial u}{\partial x} + \frac{\partial v}{\partial y} + \frac{\partial w}{\partial z}\right) + 2G\frac{\partial u}{\partial x} \qquad (5.2)$$

Normal stress component for y and z directions can be expressed in similar fashion:

$$\sigma_y = \lambda\left(\frac{\partial u}{\partial x} + \frac{\partial v}{\partial y} + \frac{\partial w}{\partial z}\right) + 2G\frac{\partial v}{\partial y} \qquad (5.3)$$

$$\sigma_z = \lambda\left(\frac{\partial u}{\partial x} + \frac{\partial v}{\partial y} + \frac{\partial w}{\partial z}\right) + 2G\frac{\partial w}{\partial z} \qquad (5.4)$$

The following is obtained by applying the relationships in Eq. (3.6) to corresponding terms in stress–strain relationships for τ_{xy} as shown in Eq. (5.1):

$$\tau_{xy} = G\left(\frac{\partial u}{\partial y} + \frac{\partial v}{\partial x}\right) \qquad (5.5)$$

Similarly, shear stress component for yz and xz planes can be expressed as follows:

$$\tau_{yz} = G\left(\frac{\partial v}{\partial z} + \frac{\partial w}{\partial y}\right) \qquad (5.6)$$

$$\tau_{xz} = G\left(\frac{\partial u}{\partial z} + \frac{\partial w}{\partial x}\right) \qquad (5.7)$$

5.3 NAVIER EQUATIONS

By substituting Eqs. (5.2), (5.5) and (5.7) into the stress equilibrium equation as shown in Eq. (2.5), we can get the following:

$$\frac{\partial\left[\lambda\left(\frac{\partial u}{\partial x} + \frac{\partial v}{\partial y} + \frac{\partial w}{\partial z}\right) + 2G\frac{\partial u}{\partial x}\right]}{\partial x} + \frac{\partial\left[G\left(\frac{\partial u}{\partial y} + \frac{\partial v}{\partial x}\right)\right]}{\partial y} + \frac{\partial\left[G\left(\frac{\partial u}{\partial z} + \frac{\partial w}{\partial x}\right)\right]}{\partial z} + f_x = 0$$

By simplifying the equation above, we get this:

$$\lambda\left(\frac{\partial^2 u}{\partial x^2} + \frac{\partial^2 v}{\partial x\partial y} + \frac{\partial^2 w}{\partial x\partial z}\right) + 2G\frac{\partial^2 u}{\partial x^2} + G\left(\frac{\partial^2 u}{\partial y^2} + \frac{\partial^2 v}{\partial x\partial y}\right) + G\left(\frac{\partial^2 u}{\partial z^2} + \frac{\partial^2 w}{\partial x\partial z}\right) + f_x$$
$$= 0$$

The following equation is obtained after expansion and rearrangement of the equation above:

$$\lambda\left(\frac{\partial^2 u}{\partial x^2} + \frac{\partial^2 v}{\partial x\partial y} + \frac{\partial^2 w}{\partial x\partial z}\right) + G\frac{\partial^2 u}{\partial x^2} + G\frac{\partial^2 v}{\partial x\partial y} + G\frac{\partial^2 w}{\partial x\partial z} + G\frac{\partial^2 u}{\partial x^2} + G\frac{\partial^2 u}{\partial y^2} + G\frac{\partial^2 u}{\partial z^2}$$
$$+ f_x = 0$$

Simplify the equation above by factorise it with common terms yields the following:

$$(\lambda + G)\left(\frac{\partial^2 u}{\partial x^2} + \frac{\partial^2 v}{\partial x\partial y} + \frac{\partial^2 w}{\partial x\partial z}\right) + G\left(\frac{\partial^2 u}{\partial x^2} + \frac{\partial^2 u}{\partial y^2} + \frac{\partial^2 u}{\partial z^2}\right) + f_x = 0$$

$$(\lambda + G)\frac{\partial}{\partial x}\left(\frac{\partial u}{\partial x} + \frac{\partial v}{\partial y} + \frac{\partial w}{\partial z}\right) + G\left(\frac{\partial^2 u}{\partial x^2} + \frac{\partial^2 u}{\partial y^2} + \frac{\partial^2 u}{\partial z^2}\right) + f_x = 0 \qquad (5.8)$$

In Eq. (2.45), hydrostatic strain can be obtained by replacing the stress components with corresponding strain components:

$$\varepsilon_m = \frac{\varepsilon_x + \varepsilon_y + \varepsilon_z}{3}$$

$$3\varepsilon_m = \varepsilon_x + \varepsilon_y + \varepsilon_z$$

By applying the relationships in Eqs. (3.2), (3.3), (3.4) to equation above yields the follows:

$$3\varepsilon_m = \frac{\partial u}{\partial x} + \frac{\partial v}{\partial y} + \frac{\partial w}{\partial z}$$

Equation below is obtained by substituting the equation above to Eq. (5.8):

$$(\lambda + G)\frac{3\partial\varepsilon_m}{\partial x} + G\left(\frac{\partial^2 u}{\partial x^2} + \frac{\partial^2 u}{\partial y^2} + \frac{\partial^2 u}{\partial z^2}\right) + f_x = 0 \qquad (5.9)$$

Laplace operation is defined as follows:

$$\nabla^2 = \frac{\partial^2}{\partial x^2} + \frac{\partial^2}{\partial y^2} + \frac{\partial^2}{\partial z^2}$$

$\frac{\partial^2 u}{\partial x^2} + \frac{\partial^2 u}{\partial y^2} + \frac{\partial^2 u}{\partial z^2}$ can be described using $\nabla^2 u$, and thus Eq. (5.9) can be simplified and written as:

$$(\lambda + G)\frac{3\partial\varepsilon_m}{\partial x} + G\nabla^2 u + f_x = 0 \qquad (5.10)$$

By substituting Eq. (4.90) into the term $\lambda + G$, we can get:

$$\lambda + G = \frac{2G\nu}{(1 - 2\nu)} + G$$

$$\lambda + G = \frac{2G\nu}{(1 - 2\nu)} + \frac{G(1 - 2\nu)}{(1 - 2\nu)}$$

$$\lambda + G = \frac{2G\nu + G - 2G\nu}{(1 - 2\nu)}$$

$$\lambda + G = \frac{G}{1 - 2\nu}$$

By substituting the relationship above into Eq. (5.10) leads to the follows:

$$\frac{G}{1 - 2\nu} \times \frac{3\partial\varepsilon_m}{\partial x} + G\nabla^2 u + f_x = 0$$
$$\frac{3G}{1 - 2\nu}\left(\frac{\partial\varepsilon_m}{\partial x}\right) + G\nabla^2 u + f_x = 0 \qquad (5.11)$$
$$\frac{3}{1 - 2\nu}\left(\frac{\partial\varepsilon_m}{\partial x}\right) + \nabla^2 u + \frac{f_x}{G} = 0$$

Similar expressions can be written for y and z directions:

$$\frac{3}{1 - 2\nu}\left(\frac{\partial\varepsilon_m}{\partial y}\right) + \nabla^2 v + \frac{f_y}{G} = 0 \qquad (5.12)$$

$$\frac{3}{1 - 2\nu}\left(\frac{\partial \varepsilon_m}{\partial z}\right) + \nabla^2 w + \frac{f_z}{G} = 0 \tag{5.13}$$

Eqs. (5.11), (5.12) and (5.13) are known as Navier equations for each axis.

5.4   BELTAMI–MICHELL STRESS COMPATIBILITY EQUATIONS

Under plane stress condition, $\sigma_z = \tau_{yz} = \tau_{xz} = 0$. Eq. (4.80) will be written as follows:

$$\varepsilon_x = \frac{1}{E}(\sigma_x - \nu\sigma_y) \tag{5.14}$$

Similarly, Eq. (4.81) will be written as below:

$$\varepsilon_y = \frac{1}{E}(\sigma_y - \nu\sigma_x) \tag{5.15}$$

From Eq. (4.94), τ_{xy} can be expressed as follows:

$$\tau_{xy} = \frac{E}{2(1 + \nu)}\gamma_{xy}$$

By expressing γ_{xy} in term of τ_{xy} yields the following equation:

$$\gamma_{xy} = \frac{2(1 + \nu)}{E}\tau_{xy} \tag{5.16}$$

Equation below is obtained by substituting the relationships in Eqs. (5.14), (5.15) and (5.16) into Eq. (3.20):

$$\frac{\partial^2}{\partial y^2}\left[\frac{1}{E}(\sigma_x - \nu\sigma_y)\right] + \frac{\partial^2}{\partial x^2}\left[\frac{1}{E}(\sigma_y - \nu\sigma_x)\right] = \frac{\partial^2}{\partial x \partial y}\left[\frac{2(1 + \nu)}{E}\tau_{xy}\right]$$

Remove the common term $\frac{1}{E}$ from both sides gives us the follows:

$$\frac{\partial^2}{\partial y^2}(\sigma_x - \nu\sigma_y) + \frac{\partial^2}{\partial x^2}(\sigma_y - \nu\sigma_x) = \frac{\partial^2}{\partial x \partial y}[2(1 + \nu)\tau_{xy}]$$

By expanding the equation above leads to this:

$$\frac{\partial^2\sigma_x}{\partial y^2} - \nu\frac{\partial^2\sigma_y}{\partial y^2} + \frac{\partial^2\sigma_y}{\partial x^2} - \nu\frac{\partial^2\sigma_x}{\partial x^2} = 2(1 + \nu)\frac{\partial^2\tau_{xy}}{\partial x \partial y} \tag{5.17}$$

Under plane stress condition, Eqs. (2.5) and (2.6) are written as follow:

$$\frac{\partial \sigma_x}{\partial x} + \frac{\partial \tau_{yx}}{\partial y} + f_x = 0 \tag{5.18}$$

$$\frac{\partial \tau_{xy}}{\partial x} + \frac{\partial \sigma_y}{\partial y} + f_y = 0 \tag{5.19}$$

Differentiate the terms in Eq. (5.18) with respect to x and we obtain this:

$$\frac{\partial^2 \sigma_x}{\partial x^2} + \frac{\partial^2 \tau_{yx}}{\partial x \partial y} + \frac{\partial f_x}{\partial x} = 0 \tag{5.20}$$

Also, differentiate the terms in Eq. (5.19) with respect to y yields the following equation:

$$\frac{\partial^2 \tau_{xy}}{\partial x \partial y} + \frac{\partial^2 \sigma_y}{\partial y^2} + \frac{\partial f_y}{\partial y} = 0 \tag{5.21}$$

By adding Eqs. (5.20) to (5.21), and noting that $\tau_{xy} = \tau_{yx}$, gives the following result:

$$\frac{\partial^2 \sigma_x}{\partial x^2} + \frac{\partial^2 \sigma_y}{\partial y^2} + 2\frac{\partial^2 \tau_{xy}}{\partial x \partial y} + \frac{\partial f_x}{\partial x} + \frac{\partial f_y}{\partial y} = 0$$

By rearranging the equation above the following is obtained:

$$\frac{\partial^2 \sigma_x}{\partial x^2} + \frac{\partial^2 \sigma_y}{\partial y^2} + \frac{\partial f_x}{\partial x} + \frac{\partial f_y}{\partial y} = -2\frac{\partial^2 \tau_{xy}}{\partial x \partial y}$$
$$-\left(\frac{\partial^2 \sigma_x}{\partial x^2} + \frac{\partial^2 \sigma_y}{\partial y^2} + \frac{\partial f_x}{\partial x} + \frac{\partial f_y}{\partial y}\right) = 2\frac{\partial^2 \tau_{xy}}{\partial x \partial y} \tag{5.22}$$

By substituting Eq. (5.22) into Eq. (5.17) yields the follows:

$$\frac{\partial^2 \sigma_x}{\partial y^2} - v\frac{\partial^2 \sigma_y}{\partial y^2} + \frac{\partial^2 \sigma_y}{\partial x^2} - v\frac{\partial^2 \sigma_x}{\partial x^2} = -(1+v)\left(\frac{\partial^2 \sigma_x}{\partial x^2} + \frac{\partial^2 \sigma_y}{\partial y^2} + \frac{\partial f_x}{\partial x} + \frac{\partial f_y}{\partial y}\right)$$

Equation below is obtained through expansion of equation above:

$$\frac{\partial^2 \sigma_x}{\partial y^2} - v\frac{\partial^2 \sigma_y}{\partial y^2} + \frac{\partial^2 \sigma_y}{\partial x^2} - v\frac{\partial^2 \sigma_x}{\partial x^2} = -\frac{\partial^2 \sigma_x}{\partial x^2} - v\frac{\partial^2 \sigma_x}{\partial x^2} - \frac{\partial^2 \sigma_y}{\partial y^2} - v\frac{\partial^2 \sigma_y}{\partial y^2} - (1+v)\left(\frac{\partial f_x}{\partial x} + \frac{\partial f_y}{\partial y}\right)$$

Eliminating the common terms on both sides yields the follows:

$$\frac{\partial^2 \sigma_x}{\partial y^2} + \frac{\partial^2 \sigma_y}{\partial x^2} = -\frac{\partial^2 \sigma_x}{\partial x^2} - \frac{\partial^2 \sigma_y}{\partial y^2} - (1 + v)\left(\frac{\partial f_x}{\partial x} + \frac{\partial f_y}{\partial y}\right)$$

Rearrange the equation above and we get:

$$\frac{\partial^2 \sigma_x}{\partial y^2} + \frac{\partial^2 \sigma_x}{\partial x^2} + \frac{\partial^2 \sigma_y}{\partial x^2} + \frac{\partial^2 \sigma_y}{\partial y^2} = -(1 + v)\left(\frac{\partial f_x}{\partial x} + \frac{\partial f_y}{\partial y}\right)$$

Through simplification of equation above, the following is obtained:

$$\left(\frac{\partial^2}{\partial x^2} + \frac{\partial^2}{\partial y^2}\right)(\sigma_x + \sigma_y) = -(1 + v)\left(\frac{\partial f_x}{\partial x} + \frac{\partial f_y}{\partial y}\right)$$

Further simplifying the equation above by introducing Laplace operator yields this:

$$\nabla^2 (\sigma_x + \sigma_y) = -(1 + v)\left(\frac{\partial f_x}{\partial x} + \frac{\partial f_y}{\partial y}\right) \tag{5.23}$$

Eq. (5.23) is the Beltrami-Michell stress compatibility equation for plane stress condition. Under the special case where body forces (f_x and f_y) are constant or zero, Eq. (5.23) will be reduced to:

$$\nabla^2 (\sigma_x + \sigma_y) = 0$$

Under plane strain condition, $\epsilon_z = \gamma_{yz} = \gamma_{xz} = 0$, and $\sigma_z \neq 0$. Eq. (4.82) will be written as follows:

$$0 = \frac{1}{E}\left[\sigma_z - v(\sigma_x + \sigma_y)\right]$$

Express the equation above with σ_z in terms of other parameters gives us the following:

$$\sigma_z = v(\sigma_x + \sigma_y) \tag{5.24}$$

By substituting Eq. (5.24) into Eq. (4.80) produces the following:

$$\varepsilon_x = \frac{1}{E}\left[\sigma_x - v\left(\sigma_y + v(\sigma_x + \sigma_y)\right)\right]$$

Simplify the equation above yields the following:

$$\varepsilon_x = \frac{1}{E}\left[\sigma_x - \nu\sigma_y - \nu^2(\sigma_x + \sigma_y)\right]$$

$$\varepsilon_x = \frac{1}{E}\left[\sigma_x - \nu\sigma_y - \nu^2\sigma_x - \nu^2\sigma_y\right] \tag{5.25}$$

$$\varepsilon_x = \frac{1}{E}\left[(1 - \nu^2)\sigma_x - \nu(1 + \nu)\sigma_y\right]$$

Similarly, for y-axis, its normal strain can be written as:

$$\varepsilon_y = \frac{1}{E}\left[(1 - \nu^2)\sigma_y - \nu(1 + \nu)\sigma_x\right] \tag{5.26}$$

The following results through substitution of Eqs. (5.25), (5.26) and (5.16) into Eq. (3.20):

$$\frac{\partial^2}{\partial y^2}\left[\frac{1}{E}[(1 - \nu^2)\sigma_x - \nu(1 + \nu)\sigma_y]\right] + \frac{\partial^2}{\partial x^2}\left[\frac{1}{E}[(1 - \nu^2)\sigma_y - \nu(1 + \nu)\sigma_x]\right]$$

$$= \frac{\partial^2}{\partial x \partial y}\left[\frac{2(1 + \nu)}{E}\tau_{xy}\right]$$

After removal of common term $\frac{1}{E}$ from both sides, we get:

$$\frac{\partial^2}{\partial y^2}\left[(1 - \nu^2)\sigma_x - \nu(1 + \nu)\sigma_y\right] + \frac{\partial^2}{\partial x^2}\left[(1 - \nu^2)\sigma_y - \nu(1 + \nu)\sigma_x\right]$$

$$= \frac{\partial^2}{\partial x \partial y}[2(1 + \nu)\tau_{xy}]$$

Expansion of equation above provides us the following equation:

$$(1 - \nu^2)\frac{\partial^2\sigma_x}{\partial y^2} - \nu(1 + \nu)\frac{\partial^2\sigma_y}{\partial y^2} + (1 - \nu^2)\frac{\partial^2\sigma_y}{\partial x^2} - \nu(1 + \nu)\frac{\partial^2\sigma_x}{\partial x^2} = 2(1 + \nu)\frac{\partial^2\tau_{xy}}{\partial x \partial y}$$

After rearranging the equation above, the following is obtained:

$$(1 - \nu^2)\left(\frac{\partial^2\sigma_x}{\partial y^2} + \frac{\partial^2\sigma_y}{\partial x^2}\right) - \nu(1 + \nu)\left(\frac{\partial^2\sigma_x}{\partial x^2} + \frac{\partial^2\sigma_y}{\partial y^2}\right) = 2(1 + \nu)\frac{\partial^2\tau_{xy}}{\partial x \partial y}$$

Eliminating the term $(1 + \nu)$ from both sides yields the following:

$$(1 - \nu)\left(\frac{\partial^2 \sigma_x}{\partial y^2} + \frac{\partial^2 \sigma_y}{\partial x^2}\right) - \nu\left(\frac{\partial^2 \sigma_x}{\partial x^2} + \frac{\partial^2 \sigma_y}{\partial y^2}\right) = 2\frac{\partial^2 \tau_{xy}}{\partial x \partial y}$$

By substituting Eq. (5.22) into the equation above we can get:

$$(1 - \nu)\left(\frac{\partial^2 \sigma_x}{\partial y^2} + \frac{\partial^2 \sigma_y}{\partial x^2}\right) - \nu\left(\frac{\partial^2 \sigma_x}{\partial x^2} + \frac{\partial^2 \sigma_y}{\partial y^2}\right) = -\left(\frac{\partial^2 \sigma_x}{\partial x^2} + \frac{\partial^2 \sigma_y}{\partial y^2} + \frac{\partial f_x}{\partial x} + \frac{\partial f_y}{\partial y}\right)$$

Expanding and rearranging the equation above gives us the following:

$$(1 - \nu)\left(\frac{\partial^2 \sigma_x}{\partial y^2} + \frac{\partial^2 \sigma_y}{\partial x^2}\right) - \nu\frac{\partial^2 \sigma_x}{\partial x^2} - \nu\frac{\partial^2 \sigma_y}{\partial y^2} = -\frac{\partial^2 \sigma_x}{\partial x^2} - \frac{\partial^2 \sigma_y}{\partial y^2} - \frac{\partial f_x}{\partial x} - \frac{\partial f_y}{\partial y}$$

$$(1 - \nu)\left(\frac{\partial^2 \sigma_x}{\partial y^2} + \frac{\partial^2 \sigma_y}{\partial x^2}\right) - \nu\frac{\partial^2 \sigma_x}{\partial x^2} - \nu\frac{\partial^2 \sigma_y}{\partial y^2} + \frac{\partial^2 \sigma_x}{\partial x^2} + \frac{\partial^2 \sigma_y}{\partial y^2} = -\frac{\partial f_x}{\partial x} - \frac{\partial f_y}{\partial y}$$

Simplification of the equation above results in this:

$$(1 - \nu)\left(\frac{\partial^2 \sigma_x}{\partial y^2} + \frac{\partial^2 \sigma_y}{\partial x^2}\right) + (1 - \nu)\left(\frac{\partial^2 \sigma_x}{\partial x^2} + \frac{\partial^2 \sigma_y}{\partial y^2}\right) = -\left(\frac{\partial f_x}{\partial x} + \frac{\partial f_y}{\partial y}\right)$$

$$\left(\frac{\partial^2 \sigma_x}{\partial y^2} + \frac{\partial^2 \sigma_y}{\partial x^2}\right) + \left(\frac{\partial^2 \sigma_x}{\partial x^2} + \frac{\partial^2 \sigma_y}{\partial y^2}\right) = -\frac{1}{1 - \nu}\left(\frac{\partial f_x}{\partial x} + \frac{\partial f_y}{\partial y}\right)$$

$$\left(\frac{\partial^2}{\partial x^2} + \frac{\partial^2}{\partial y^2}\right)(\sigma_x + \sigma_y) = -\frac{1}{1 - \nu}\left(\frac{\partial f_x}{\partial x} + \frac{\partial f_y}{\partial y}\right)$$

Further simplifying the equation above by introducing Laplace operator yields the following:

$$\nabla^2(\sigma_x + \sigma_y) = -\frac{1}{1 - \nu}\left(\frac{\partial f_x}{\partial x} + \frac{\partial f_y}{\partial y}\right) \tag{5.27}$$

Eq. (5.27) is the Beltrami-Michell stress compatibility equation for plane strain condition. Under the special case where body forces (f_x and f_y) are constant or zero, Eq. (5.27) will be reduced to:

$$\nabla^2(\sigma_x + \sigma_y) = 0$$

5.5 AIRY STRESS FUNCTION

When at rest, a body possesses potential energy, say ψ. Body force can thus be expressed in term of such potential energy:

$$f_x = -\frac{\partial \psi}{\partial x} f_y = -\frac{\partial \psi}{\partial y} f_z = -\frac{\partial \psi}{\partial z} \tag{5.28}$$

Also, the stress component can be expressed in term of stress function, ϕ. For a 2-D scenario,

$$\sigma_x = \frac{\partial^2 \phi}{\partial y^2} + \psi \sigma_y = \frac{\partial^2 \phi}{\partial x^2} + \psi \tau_{xy} = -\frac{\partial^2 \phi}{\partial x \partial y} \tag{5.29}$$

Under plane stress condition, the equilibrium equation can be written as follows by substituting relationships in Eq. (5.28) into Eq. (5.18):

$$\frac{\partial \sigma_x}{\partial x} + \frac{\partial \tau_{yx}}{\partial y} - \frac{\partial \psi}{\partial x} = 0$$

Simplify the equation above yields the following equation:

$$\frac{\partial}{\partial x}(\sigma_x - \psi) + \frac{\partial \tau_{yx}}{\partial y} = 0 \tag{5.30}$$

Similarly, the following equation can be derived from Eq. (5.19):

$$\frac{\partial}{\partial y}(\sigma_y - \psi) + \frac{\partial \tau_{xy}}{\partial x} = 0 \tag{5.31}$$

We get the equation below by substituting relationships in Eqs. (5.28) and (5.29) into Eq. (5.23):

$$\nabla^2\left(\frac{\partial^2 \phi}{\partial y^2} + \psi + \frac{\partial^2 \phi}{\partial x^2} + \psi\right) = -(1 + v)\left[\frac{\partial}{\partial x}\left(-\frac{\partial \psi}{\partial x}\right) + \frac{\partial}{\partial y}\left(-\frac{\partial \psi}{\partial y}\right)\right]$$

With simplification, the equation below is obtained:

$$\nabla^2\left(\frac{\partial^2\phi}{\partial x^2} + \frac{\partial^2\phi}{\partial y^2} + 2\psi\right) = (1 + \nu)\left(\frac{\partial^2\psi}{\partial x^2} + \frac{\partial^2\psi}{\partial y^2}\right)$$

By introducing Laplace operator to the equation, it can be transformed to this form:

$$\nabla^2(\nabla^2\phi + 2\psi) = (1 + \nu)(\nabla^2\psi)$$

Expand the equation above gives us the follows:

$$\nabla^4\phi + 2\nabla^2\psi = \nabla^2\psi + \nu\nabla^2\psi$$

By rearranging the equation above leads to the follows:

$$\nabla^4\phi + 2\nabla^2\psi - \nabla^2\psi - \nu\nabla^2\psi = 0$$

$$\nabla^4\phi + \nabla^2\psi - \nu\nabla^2\psi = 0$$

Simplify the equation above yields the follows:

$$\nabla^4\phi + (1 - \nu)\nabla^2\psi = 0 \qquad (5.32)$$

Eq. (5.32) is a biharmonic equation for plane stress condition. Under a special case where no body force presents, Eq. (5.32) will be reduced to:

$$\nabla^4\phi = 0$$

Under plane stress condition, the equilibrium equation can be written as follows by substituting relationships in Eqs. (5.28) and (5.29) into Eq. (5.27):

$$\nabla^2\left(\frac{\partial^2\phi}{\partial y^2} + \psi + \frac{\partial^2\phi}{\partial x^2} + \psi\right) = -\frac{1}{1 - \nu}\left[\frac{\partial}{\partial x}\left(-\frac{\partial\psi}{\partial x}\right) + \frac{\partial}{\partial y}\left(-\frac{\partial\psi}{\partial y}\right)\right]$$

Simplify the equation above gives us:

$$\nabla^2\left(\frac{\partial^2\phi}{\partial x^2} + \frac{\partial^2\phi}{\partial y^2} 2\psi\right) = \frac{1}{1 - \nu}\left(\frac{\partial^2\psi}{\partial x^2} + \frac{\partial^2\psi}{\partial y^2}\right)$$

By introducing the Laplace operator to the equation leads to the follows:

$$\nabla^2(\nabla^2\phi + 2\psi) = \frac{1}{1-\nu}(\nabla^2\psi)$$

Expand the equation above yields the following equation:

$$\nabla^4\phi + 2\nabla^2\psi = \frac{\nabla^2\psi}{1-\nu}$$

Rearrange the equation above results in the follows:

$$\nabla^4\phi + 2\nabla^2\psi - \frac{\nabla^2\psi}{1-\nu} = 0$$

$$\nabla^4\phi + \frac{2\nabla^2\psi(1-\nu)}{1-\nu} - \frac{\nabla^2\psi}{1-\nu} = 0$$

$$\nabla^4\phi + \frac{2\nabla^2\psi - 2\nu\nabla^2\psi - \nabla^2\psi}{1-\nu} = 0$$

$$\nabla^4\phi + \frac{\nabla^2\psi - 2\nu\nabla^2\psi}{1-\nu} = 0$$

Simplify the equation above gives us the following expression:

$$\nabla^4\phi + \frac{1-2\nu}{1-\nu}\nabla^2\psi = 0 \tag{5.33}$$

Eq. (5.33) is a biharmonic equation for plane strain condition. Under a special case where no body force presents, Eq. (5.33) will be reduced to:

$$\nabla^4\phi = 0$$

6 Solutions for Plasticity

6.1 INTRODUCTION

There is no mean to determine the plastic behaviour of a material, unless through experiments. Similarly, structural design that considers material plasticity involves complex calculations and iterations as the gradient of curve (yields the amount of stress required to produce a unit of strain) varies from one point to another. Engineers usually adopt linear plastic curve, which is easy to implement yet conservative enough to be safe.

By considering plastic behaviour, an element is expected to carry more stress before the plastic hinge formed (fracture). In this way, element design from elastic design can be optimised. In structural engineering, plasticity is allowed in structural steel design, due to the property of steel where strain hardening will occur at some point beyond its elastic limit. This is a favourable situation that enable steel to carry load without failing drastically, and safe enough for an engineer to allow plastic behaviour in steel. In Eurocode 3, plasticity is allowed only for class 1 and 2 section, where local buckling will not occur within the member and causes premature loss of sectional resistance.

For plasticity analysis, yield point is important: it is the point where the material changes from having linear stress–strain relationship (elastic) to non-linear (plastic). To ease the analysis, mathematical models are developed alongside simplification of a stress–strain diagram. Fig. 6.1 below shows such effort for perfect plastic and plastic with linear work hardening behaviours.

6.2 YIELDS CRITERIA

For material scientists and engineers, the amount of stress a material can sustain before it exhibits plastic behaviour is concerned. Given the stress–strain diagram for the material, different yield criteria can be applied depending on the actual situation. Despite the differences, the concepts of those criteria are similar: if the stress developed within a body exceeds its yield strength, then the body is expected to exhibit plastic behaviour. To date, yield criterion always assumes the material to be isotropic. For other types of material, the plastic behaviour will be too difficult and complicated to predict.

6.2.1 TRESCA YIELDS CRITERION

For Tresca yield criterion, yielding is initiated when the maximum shear stress, i.e. the largest of the three developed maximum shear stresses, exceeds the threshold value of material.

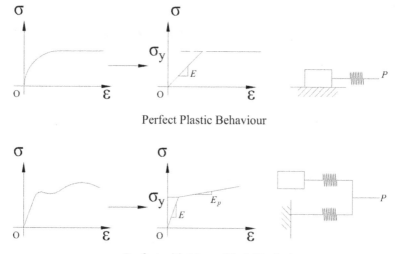

FIGURE 6.1   Perfect plastic and plastic with linear work hardening.

Let define maximum shear stress as it is in Eq. (2.44), where σ_1 and σ_3 are the maximum and minimum principal stresses respectively. Also, let k be the threshold value that defines whether the developed shear stress is enough to make the body initiate yielding. Thus, at yielding point:

$$\frac{(\sigma_1 - \sigma_3)}{2} = k \qquad\qquad (6.1)$$

Consider the case where only tensile stress instead of shear stress developed in the body. Express Eq. (6.1) in the following form:

$$\sigma_1 - \sigma_3 = 2k$$

The yield stress of material, σ_y is determined through uniaxial load test. Therefore, σ_y is the threshold value that defines whether the developed tensile stress is enough to make the body initiate yielding. For this case, σ_1 should equal to the yield strength of material, σ_y, while principal stress along other directions, σ_2 and σ_3 are zero at yielding point. By applying the condition to equation above yields the follows:

$$\sigma_y = 2k$$

Express k in term of σ_y produces the expression below:

$$k = \frac{\sigma_y}{2}$$

By Tresca yield criterion, the stress required for a body to yield in shear is half the amount of that required for a body to yield in tension.

6.2.2 VON MISES YIELDS CRITERION

Von Mises yield criterion is devised based on distortion energy theory. According to this criterion, yielding is initiated when the second deviatoric invariant exceeds the threshold value of material.

The equation below is obtained by substituting the expressions of I_1 and I_2 (Eqs. (2.35) and (2.36)) into J_2 (Eq. (2.47)):

$$J_2 = \frac{(\sigma_x + \sigma_y + \sigma_z)^2}{3} - (\sigma_x \sigma_y + \sigma_y \sigma_z + \sigma_x \sigma_z - \tau_{xy}^2 - \tau_{yz}^2 - \tau_{xz}^2)$$

After expanding the equation above it becomes:

$$J_2 = \frac{\sigma_x^2 + \sigma_x \sigma_y + \sigma_x \sigma_z + \sigma_y \sigma_x + \sigma_y^2 + \sigma_y \sigma_z + \sigma_z \sigma_x + \sigma_z \sigma_y + \sigma_z^2}{3}$$
$$- \sigma_x \sigma_y - \sigma_y \sigma_z - \sigma_x \sigma_z + \tau_{xy}^2 + \tau_{yz}^2 + \tau_{xz}^2$$

Simplify the equation above yields the follows:

$$J_2 = \frac{\sigma_x^2 + \sigma_y^2 + \sigma_z^2}{3} - \frac{1}{3}\sigma_x \sigma_y - \frac{1}{3}\sigma_y \sigma_z - \frac{1}{3}\sigma_x \sigma_z + \tau_{xy}^2 + \tau_{yz}^2 + \tau_{xz}^2$$

Express the equation above in terms of principal stresses, where $\sigma_x = \sigma_1$, $\sigma_y = \sigma_2$, $\sigma_z = \sigma_3$ and $\tau_{xy} = \tau_{yz} = \tau_{xz} = 0$ and we get:

$$J_2 = \frac{\sigma_1^2 + \sigma_2^2 + \sigma_3^2}{3} - \frac{1}{3}\sigma_1 \sigma_2 - \frac{1}{3}\sigma_2 \sigma_3 - \frac{1}{3}\sigma_1 \sigma_3$$

Factorization of the equation above with $\frac{1}{6}$ gives us:

$$J_2 = \frac{1}{6}(2\sigma_1^2 + 2\sigma_2^2 + 2\sigma_3^2 - 2\sigma_1 \sigma_2 - 2\sigma_2 \sigma_3 - 2\sigma_1 \sigma_3)$$

Rearrange the equation above yields the follows:

$$J_2 = \frac{1}{6}[(\sigma_1^2 - 2\sigma_1 \sigma_2 + \sigma_2^2) + (\sigma_2^2 - 2\sigma_2 \sigma_3 + \sigma_3^2) + (\sigma_3^2 - 2\sigma_1 \sigma_3 + \sigma_1^2)]$$

Factorise each grouped quadratic equation leads to this:

$$J_2 = \frac{1}{6}[(\sigma_1 - \sigma_2)^2 + (\sigma_2 - \sigma_3)^2 + (\sigma_3 - \sigma_1)^2] \tag{6.2}$$

Let k be the threshold value that defines whether the developed shear stress is enough to make the body initiate yielding. Also, let σ_1 and σ_2 be the coupled shear stresses that achieve threshold value:

$$\sigma_1 = -\sigma_2 = k \quad \sigma_3 = 0 \tag{6.3}$$

At yielding point, by applying the relationship in Eqs. (6.3) to (6.2) becomes:

$$J_2 = \frac{1}{6}[(k - (-k))^2 + ((-k) - 0)^2 + (0 - k)^2]$$

Simplify the equation above and we get:

$$J_2 = \frac{1}{6}[(2k)^2 + (-k)^2 + (-k)^2]$$
$$J_2 = k^2 \tag{6.4}$$

σ_y is the threshold value that defines whether the developed tensile stress is enough to make the body initiate yielding. Let σ_1 be the yield stress of the body and $\sigma_2 = \sigma_3 = 0$, Eq. (6.2) becomes:

$$J_2 = \frac{1}{6}\left[\left(\sigma_y - 0\right)^2 + (0 - 0)^2 + \left(0 - \sigma_y\right)^2\right]$$

By simplifying the equation above we obtain:

$$J_2 = \frac{1}{6}\left[(\sigma_y)^2 + (-\sigma_y)^2\right]$$
$$J_2 = \frac{1}{6}\left[2\sigma_y^2\right] \tag{6.5}$$
$$J_2 = \frac{\sigma_y^2}{3}$$

Equalise the Eqs. (6.4) and (6.5) yields the follows:

$$k^2 = \frac{\sigma_y^2}{3}$$

Square root both sides in the equation above results in the following expression:

$$k = \frac{\sigma_y}{\sqrt{3}}$$

By Von Mises yield criterion, the stress required for a body to yield in shear is $\frac{1}{\sqrt{3}}$ times of the amount required for a body to yield in tension.

6.3   PLASTIC WORK

Work done is the product of force and displacement. By this definition:

$$W = F \times dx$$

Plastic work is done by external force when permanent deformation is formed in a body, as shown in Fig. 6.2.

Work done per unit volume of such body can be expressed as:

$$dW = \frac{F \times dx}{V}$$
$$dW = \frac{F \times dx}{A \times L}$$
$$dW = \sigma \times d\varepsilon$$

In matrix form, the equation above is expressed in:

$$dW = \{d\varepsilon\}^T \{\sigma\}$$

Permanent deformation caused in plastic state is the summation of elastic and plastic deformations. Therefore, the strain, $d\varepsilon$ can be divided into two components: elastic strain, $d\varepsilon_e$ and plastic strain, $d\varepsilon_p$. Equation above can thus be written as follows:

$$dW = \{d\varepsilon_e\}^T \{\sigma\} + \{d\varepsilon_p\}^T \{\sigma\}$$

The work done related to elastic strain is reversible as the body will restore back to its original shape upon unloading. Therefore, the plastic work is solely due to plastic

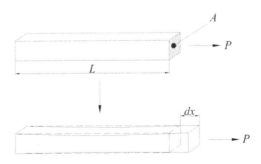

FIGURE 6.2   Plastic work in a body.

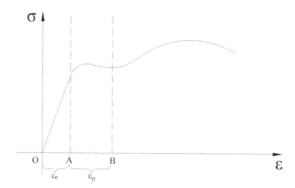

FIGURE 6.3 Definition of plastic work in stress–strain curve.

strain. The area of graph starting from point A to B under the stress–strain curve denotes the plastic work as shown in Fig. 6.3.

Mathematically, plastic work can be defined as:

$$dW = \int \{d\epsilon_p\}^T \{\sigma\}$$

6.4 ASSOCIATED FLOW RULE

Similar to Hooke's law, plastic behaviour is also governed by its constitutive equations. This rule is known as flow rule. In elastic state, the strains are linearly depending on the state of stress in a body. In plastic state however, this is not the case. The strains are governed by stress and loading history or the body, as the stress–strain relationship of material in plastic zone will be altered from time to time due to the break of bond between particles. To take this condition into account, incremental analysis is required. This method will first set up the timeframe for analysis, and then for each time step, the corresponding stress–strain relationship and resultant deformation will be determined. Through this method, plastic strains for each incremental loading are determined, and the total plastic strain can be derived thereafter.

The strain increment in plastic state is solely depending on the current state of stress, which can also be mathematically defined as:

$$\{d\epsilon_p\} = d\lambda \left\{ \frac{\partial f}{\partial \{\sigma\}} \right\} \tag{6.6}$$

$d\lambda$ is a non-negative constant for plasticity.

Under elastic state, stress is expressed as below:

$$\{\sigma\} = \{D\} \{\epsilon_e\}$$

Express elastic strain in term of state of stress yields the following expression:

$$\{\epsilon_e\} = \{D\}^{-1}\{\sigma\}$$

The increment of elastic strain can thus be written as follows:

$$\{d\epsilon_e\} = \{D\}^{-1}\{d\sigma\} \tag{6.7}$$

Total strain is the summation of elastic and plastic strains. From Eqs. (6.6) and (6.7), the following elastic–plastic stress–strain relationship yielded:

$$\{d\epsilon\} = \{D\}^{-1}\{d\sigma\} + d\lambda \left\{ \frac{\partial f}{\partial\{\sigma\}} \right\}$$

By expressing the elastic–plastic stress–strain relationship in matrix form yields the following expression:

$$
\begin{bmatrix}
d\epsilon_x \\
d\epsilon_y \\
d\epsilon_z \\
d\gamma_{xy} \\
d\gamma_{yz} \\
d\gamma_{zx} \\
0
\end{bmatrix}
=
\begin{bmatrix}
 & & & D^{-1} & & & \partial f/\partial\sigma_x \\
 & & & & & & \partial f/\partial\sigma_y \\
 & & & & & & \partial f/\partial\sigma_z \\
 & & & & & & \partial f/\partial\tau_{xy} \\
\partial f/\partial\sigma_x & \partial f/\partial\sigma_y & \partial f/\partial\sigma_z & \partial f/\partial\tau_{xy} & \partial f/\partial\tau_{yz} & \partial f/\partial\tau_{zx} & \partial f/\partial\tau_{yz} \\
 & & & & & & \partial f/\partial\tau_{zx} \\
 & & & & & & -a
\end{bmatrix}
\begin{bmatrix}
d\sigma_x \\
d\sigma_y \\
d\sigma_z \\
d\tau_{xy} \\
d\tau_{yz} \\
d\tau_{zx} \\
d\lambda
\end{bmatrix}
$$

After simplification we get this:

$$\left\{ \begin{matrix} d\epsilon \\ 0 \end{matrix} \right\} = \left\{ \begin{matrix} D^{-1} & A \\ A^T & -a \end{matrix} \right\} \left\{ \begin{matrix} d\sigma \\ d\lambda \end{matrix} \right\} \tag{6.8}$$

Where,

$$
[d\epsilon] =
\begin{bmatrix}
d\epsilon_x \\
d\epsilon_y \\
d\epsilon_z \\
d\gamma_{xy} \\
d\gamma_{yz} \\
d\gamma_{zx}
\end{bmatrix}, \quad
[d\sigma] =
\begin{bmatrix}
d\sigma_x \\
d\sigma_y \\
d\sigma_z \\
d\tau_{xy} \\
d\tau_{yz} \\
d\tau_{zx}
\end{bmatrix}, \quad
[A] =
\begin{bmatrix}
\partial f/\partial\sigma_x \\
\partial f/\partial\sigma_y \\
\partial f/\partial\sigma_z \\
\partial f/\partial\tau_{xy} \\
\partial f/\partial\tau_{yz} \\
\partial f/\partial\tau_{zx}
\end{bmatrix} \tag{6.9}
$$

$$[A]^T = \begin{bmatrix} \partial f/\partial\sigma_x & \partial f/\partial\sigma_y & \partial f/\partial\sigma_z & \partial f/\partial\tau_{xy} & \partial f/\partial\tau_{yz} & \partial f/\partial\tau_{zx} \end{bmatrix}$$

6.5 HARDENING EFFECT

Hardening is described using the function below:

$$f\left(\sigma, \epsilon_p, k\right)$$

In equation above, k is introduced as hardening coefficient. By differentiating the equation above yields the follows:

$$df = \left\{\frac{\partial f}{\partial \sigma}\right\}^T \{d\sigma\} + \left\{\frac{\partial f}{\partial \epsilon_p}\right\}^T \{d\epsilon_p\} + \left\{\frac{\partial f}{\partial k}\right\}^T \{dk\} \qquad (6.10)$$

The yield criterion for a work-hardening material is:

$$f\left(\sigma, \epsilon_p, k\right) = 0$$

Therefore, by equalizing Eq. (6.10) to zero results to the expression below:

$$\left\{\frac{\partial f}{\partial \sigma}\right\}^T \{d\sigma\} + \left\{\frac{\partial f}{\partial \epsilon_p}\right\}^T \left\{d\epsilon_p\right\} + \left\{\frac{\partial f}{\partial k}\right\}^T \{dk\} = 0$$

By expanding the term $\left\{\frac{\partial f}{\partial \sigma}\right\}^T$ in equation above yields the follows:

$$\left[\partial f/\partial \sigma_x \ \ \partial f/\partial \sigma_y \ \ \partial f/\partial \sigma_z \partial f/\partial \tau_{xy} \ \ \partial f/\partial \tau_{yz} \ \ \partial f/\partial \tau_{zx}\right] \{d\sigma\}$$

$$+ \left\{\frac{\partial f}{\partial \epsilon_p}\right\}^T \left\{d\epsilon_p\right\} + \left\{\frac{\partial f}{\partial k}\right\}^T \{dk\} = 0$$

By substituting the relationships in Eq. (6.9) we can obtain this:

$$[A]^T \{d\sigma\} + \left\{\frac{\partial f}{\partial \epsilon_p}\right\}^T \left\{d\epsilon_p\right\} + \left\{\frac{\partial f}{\partial k}\right\}^T \{dk\} = 0 \qquad (6.11)$$

From Eq. (6.8), the following expression obtained:

$$0 = [A]^T d\sigma - a d\lambda$$

Comparing the equation above with Eq. (6.11) yields the following relationships:

$$
-ad\lambda = \left\{\frac{\partial f}{\partial \varepsilon_p}\right\}^T \left\{d\varepsilon_p\right\} + \left\{\frac{\partial f}{\partial k}\right\}^T \{dk\}a
$$

$$
= -\frac{1}{\lambda}\left[\left\{\frac{\partial f}{\partial \varepsilon_p}\right\}^T \left\{d\varepsilon_p\right\} + \left\{\frac{\partial f}{\partial k}\right\}^T \{dk\}\right]
$$

Index